Fixed Biological Surfaces — Wastewater Treatment

The Rotating Biological Contactor

Author:

Ronald L. Antonie

Manager, Technical Services
Autotrol Corporation

Published by

CRC PRESS, Inc.
18901 Cranwood Parkway · Cleveland, Ohio 44128

Library of Congress Cataloging in Publication Data

Antonie, Ronald L.
    Fixed biological surfaces.

    At head of title: CRC.
    Bibliography: p.
    Includes index.
    1. Sewage – Purification – Biological treatment.
I. Chemical Rubber Company, Cleveland. II. Title.
TD755.A67   628′.354   76-3546
ISBN 0-87819-079-1

© 1976 by CRC Press, Inc.

International Standard Book Number 0-87819-079-1
Library of Congress Card Number 76-3546

Printed in the United States

FOREWORD

This first volume in the Uniscience Series on Water Pollution Control Technology is a concise and comprehensive state-of-the-art report on the rotating biological contactor process, an innovative development in fixed-film biological treatment of wastewaters. The author, himself one of the principals in the development and commercialization of the process, has culled and critically reviewed data from numerous technical reports, journal articles, and unpublished conference presentations.

Our objective in this book and in subsequent volumes of the series is to provide a reference manual for design engineers, planners, and managers in industry and government. This is particularly important in the present critical period for implementation of water pollution controls.

R. Prober
Case Western Reserve University
Cleveland, Ohio
December 22, 1975

PREFACE

Fixed-film biological treatment was the first controlled mechanical means employed to provide secondary treatment to domestic wastes. Because of its simplicity, it received widespread use during the first half of the 20th century. However, as urban areas increased in size and as treatment requirements became more stringent, use of suspended growth systems became more common because of lower construction costs and higher treatment capabilities. Suspended growth systems have since become the dominant type of treatment at the expense of treatment plant complexity and high energy consumption. As treatment requirements continue to become more stringent and treatment plant operation becomes more complex, and because energy costs have begun to rise, there is now a trend back to the simplicity and low energy consumption of fixed-film treatment. This trend has been accelerated by the rapid development of the rotating biological contactor (RBC), which retains the simplicity and low energy consumption of the trickling filter, but is capable of producing the same quality of effluent as suspended growth systems.

The objective of this book is to provide concise, comprehensive, and up-to-date information on the state of the art of fixed-film biological treatment and in particular on the rotating biological contactor. The material presented is not merely a compilation of papers, but is a fully integrated and coordinated presentation of all important information presently available. In addition to providing general information on the rotating biological contactor process, it is meant to serve as a guide for treatment plant design for practicing consulting engineers and regulatory authorities. It is also intended for use by planners and management in industry and government who are charged with establishing wastewater treatment objectives and evaluating the impact of the costs required to meet these objectives. Information is also provided which will be of help and interest to the treatment plant operator.

The approach taken in presenting information is principally empirical. Theoretical models were not developed to predict performance or operating principles of the process. In this sense, the book is not meant to be used as a textbook but can be used to supplement the materials presented in standard textbooks for sanitary engineering.

This book historically traces the development and use of fixed-film treatment systems, emphasizing the rotating biological contactor process. Process development and design criteria for the rotating biological contactor are presented for treatment of domestic and a wide range of industrial wastes. The recent and rapidly spreading requirements for nitrification and denitrification are also discussed along with design criteria to meet these requirements. Continued urban growth and increasing water quality standards have left many existing primary and secondary treatment plants in need of upgrading to greater hydraulic capacity and higher levels of treatment. The rotating biological contactor process can be easily incorporated into existing treatment plants for upgrading. Design criteria for this application are discussed in detail.

Recent operating experience with rotating biological contactor treatment of domestic and industrial wastes is discussed. Current capital and operating costs are also presented, which will permit comparison with alternative systems of treatment.

I wish to acknowledge the contributions of my colleagues and other researchers in the field who have assisted in developing the information presented in this book.

R. L. Antonie
Autotrol Corporation
Milwaukee, Wisconsin
November 17, 1975

THE AUTHOR

Ronald L. Antonie is Manager of Technical Services at Autotrol Corporation, Milwaukee, Wisconsin.

Mr. Antonie received his B.S. degree in Chemical Engineering in 1964 from the University of Wisconsin at Madison and received his M.B.A. degree in 1970 from the University of Wisconsin at Milwaukee. He is a member of the American Institute of Chemical Engineers and the Water Pollution Control Federation.

He has written numerous articles which pertain to the research, design, and application of the Rotating Biological Contactor Process and holds several patents concerned with rotating disc systems for chemical processing and wastewater treatment applications.

TABLE OF CONTENTS

This book is dedicated to the spirit of innovation and persistence in effort that exists among my colleagues at Autotrol Corporation.

Chapter 1

FIXED-FILM BIOLOGICAL REACTOR — THE TRICKLING FILTER

INTRODUCTION

While this book deals primarily with the rotating biological contactor (RBC) process, this first chapter is devoted to a discussion of the trickling filter process for the purpose of providing an overall background on the history and development of fixed-film biological treatment systems.

HISTORY AND FUNDAMENTALS

The trickling filter process consists of a bed of rocks or synthetic media over which wastewater is uniformly distributed. After a short period of operation, a biological growth develops. The attached microorganisms remove suspended and dissolved organic material from the wastewater as it flows over the surfaces. At the same time, oxygen is absorbed from the air by the falling film of wastewater. The natural draft created causes air to flow upward through the media at the same time the wastewater passes downward.

The attached film of microorganisms is often defined to contain two distinctly different regions of activity. The effective film depth is that which is in immediate contact with the wastewater. In this region, dissolved oxygen and organic substrates are made readily available to the organisms by mixing and diffusion from the wastewater. The area between the filter media and the effective film depth principally consists of anaerobic organisms, because all oxygen is consumed before it can diffuse to the inner layer. The relative sizes of these two films depend on the concentration of oxygen in the wastewater film and the turbulence between the water film and the attached growth. One of the principal reasons for improved trickling filter operation with effluent recycle is the increase in the relative size of the effective film depth through increased oxygen concentration in the water film and increased turbulence with the attached growth.

Excess biological growth sloughs from the filter media and is carried out of the process by the wastewater flow. One of the principal mechanisms of sloughing in the trickling filter is the anaerobic activity at the interface with the filter media. Gas production together with loss of adhesion cause relatively large amounts of biomass to be stripped from the media surface in irregular intervals. Sometimes this irregular solids production is associated with organic overloads or changes in climatic conditions.

The first controlled use of fixed-film biological treatment of wastes occurred in England in the mid-19th century where wastewater was intermittently distributed over beds of sand.[1] The principal function of the sand was to filter particulate matter from the wastewater, but it also served as a medium for attachment of microbial organisms which removed much of the nonfilterable organic matter. As the sand was later replaced by stone and an underdrain system was included, the treatment efficiency was improved. The wastewater then trickled down over bacteria-covered stones, and air passed upward to provide a continuous supply of oxygen to encourage aerobic biological activity. This system became known as the percolating filter in England and the trickling filter in the U.S. The name trickling filter has persisted to the present time even though filtration is no longer a mechanism of treatment in the process.

The first experimental use of the process in the U.S. was by the Massachusetts State Board of Health in 1889 where gravel was substituted for sand.[2] Many of the early developments, including a spray system to distribute the wastewater and an underdrain system, occurred in England toward the end of the century. Commercial use of the process began in England just after the turn of the century and in the U.S. in 1908 when plants were constructed for the cities of Columbus, Ohio, Reading, Pennsylvania, and Washington, Pennsylvania. Use of the trickling filter increased steadily from that time.

During the 1930s the high rate trickling filter process utilizing effluent recirculation was developed. It permitted greater treatment capacities for the trickling filter. From that time until the 1950s, no further developments occurred. The activated sludge process was developed in the 1920s and gradually began to gain predominance over the trickling filter process. Trickling filters were in use in 3,506 municipal wastewater treatment plants in the U.S. in 1962 as reported in a statistical summary issued by the U.S. Public Health Service in 1964. Of these plants, 2,135 employed standard filters, while 1,371 employed high-rate filters. The total population served by these facilities at that time was 23 million. For more details on the historical development of the trickling filter process, refer to Chase.[3] A more recent report[4] indicates that the total number of trickling filter plants is about 3,400, which serve about 30 million people. Apparently a number of the plants were expanded (or gradually overloaded) to serve more people, however, older plants were abandoned or converted to other treatment processes at a greater rate than new plants were built.

For further detail on the construction and operation of trickling filters, refer to Reference 2.

PROBLEM AREAS

The trickling filter process became readily accepted and widely used because of its simplicity and low operating costs, however, a number of economic and operating problems developed. The proper grade of stone could not always be found within a reasonable distance from the plant site, which significantly increased transportation costs. The type of stone often specified was "20-cycle" stone, which referred to the number of cycles of freezing and thawing the stone could withstand without deterioration. This specification was meant to provide stone which would last for 20 winter seasons, but in recent years, this specification has become very difficult to meet with reasonably available materials.

Problems of clogging by excess biomass have been experienced when using a trickling filter. This is caused by having too small an interstitial volume within the stones due to either too small a stone size or deterioration of the stones over a period of time. When clogging occurs, the wastewater channels through the media with little treatment. The clogged areas become anaerobic, generating objectionable odors, and are difficult to clear once they become clogged. Certain species of flies often breed in a trickling filter to cause a further nuisance.

Treatment efficiencies of a trickling filter decrease during winter operation because of excessive cooling of the wastewater and ice formation on the surface of the stones. Covering the trickling filter alleviates this only partially.

The principal reason for the gradual loss of popularity of the trickling filter is the limited degree of treatment achievable. The short wastewater retention time limits soluble BOD removal to the extent that it cannot meet the levels of treatment possible in an activated sludge system with a much longer retention time. With effluent discharge requirements becoming more stringent, the trickling filter could no longer compete economically with the activated sludge process.

SYNTHETIC MEDIA

In the mid-1950s, a synthetic plastic media was developed which promised to solve many of the problems of the trickling filter. The initial work reported by Dow Chemical Company in 1955 evaluated a modified berl saddle design and a corrugated sheet construction of polystyrene. They subsequently marketed the corrugated sheet design, then of PVC, under the trade name Dow Pac®. Since that time, several U.S. and European firms have begun marketing various designs of plastic media. One U.S. firm has also recently started marketing a media constructed of horizontal wooden slats. Table 1-1 compares some of the characteristics of these synthetic media.

The corrugated plastic media were widely accepted as a substitute for rock and consequently, interest was revived in the trickling filter process in the late 1950s and early 1960s. The wider openings it provided reduced the danger of plugging, which alleviated many of the problems associated with plugging of the stone media. Its lower weight allowed the media to be stacked to greater heights and reduce construction costs and land requirements albeit at the increased cost of pumping wastewater and effluent recycle to greater heights.

Several companies who supply plastic tower packings in the form of Pall rings to the chemical processing industry have recently begun selling them as a trickling filter media. While in principle this form of synthetic media will work, it still has the major disadvantages of the rock media, i.e., small interstitial openings with their tendency to plug and limited depths of packing because of the inability to support the increased weight when covered with biomass. Because of this, they have received limited use.

Although some of the largest trickling filter

TABLE 1-1

Types of Synthetic Media

Supplier	Trade name	Construction	Specific surface area, ft² /ft³
Envirotech Corp. Brisbane, CA	Surfpac*	Flat and corrugated PVC sheets	27
B. F. Goodrich Marietta, OH	Koroseal Vinyl Core	Flat and corrugated PVC sheets	30.5
ICI Great Britain	Flocor**	Flat and corrugated PVC sheets	29
Neptune-Microfloc Corvallis, OR	Del-Pak***	Horizontal wooden slats	14
Koch Eng. Co. New York, NY	Flexirings	Plastic pall rings	28
Norton Chemical Co. Akron, OH	Actifil	Plastic pall rings	29
Institute de Reserche Chimique Applique, France	Cloisonyle	PVC tubes	68.5

*Formerly available from Dow Chemical Co., Midland, MI.
**Formerly available from Ethyl Corp., Baton Rouge, LA.
***Formerly available from Del-Pak Corp., Corvallis, OR.

plants have been built in recent years, the use of the trickling filter is steadily decreasing. This has resulted from the inability of the trickling filter to consistently achieve high degrees of soluble BOD removal while effluent standards have become increasingly more stringent. Because of this, some state regulatory agencies no longer allow trickling filters to be used as the sole means of secondary treatment.

In the past, an existing rock trickling filter plant was often upgraded with synthetic media. Now with more stringent effluent standards, a different treatment process must often be substituted to assure that these standards will be consistently met. As an attempt to solve this problem, one synthetic media manufacturer (Del-Pak) has developed a system where the trickling filter is followed by an activated sludge system to improve overall treatment levels. Settled sludge from the clarifier following the aeration tank is recycled to the tower of media.

CONTACT AERATION

Contact aeration is the generic term often applied to a treatment process using closely spaced, stationary, vertical plates, which are completely immersed in wastewater. Air is introduced beneath the plates to aerate the wastewater and help strip the excess biomass which develops on the plates. This process became known in the U.S. as the "Hays Process" and in Europe as the "Emscher Filter."

The history of this fixed-film process dates back to the turn of the century in the U.S. and in England with the work of Travis, who attempted to remove colloidal matter from wastewater by immersing wooden slats in a settling tank.[5] Aeration was later added by Buswell[5] in the U.S. and by Bach[5] in Germany in the 1920s. The most significant process development work was done by Hays in the late 1930s at Waco, Texas. This work resulted in the installation of about 12 small contact aeration plants in New Jersey in the early 1940s. While these plants demonstrated that adequate secondary treatment could be obtained, they were plagued with operating problems resulting from clogged contact plates and air diffusers. These operating problems along with the growing acceptance of the activated sludge process caused a rapid demise of the contact aeration process, although small commercial units were available as recently as 1963.

Recent research work has been done on this process by Haug and McCarty[6] using a packed bed of stone media with both air and pure oxygen for aeration, but no commercial applications are known. A variation on this process using a fluidized bed of sand with attached biological growth and aeration with air or pure oxygen has recently been reported by Jeris.[7] No commercial applications are known for this system.

DESIGN CRITERIA

Trickling filter design criteria have been thoroughly discussed in sanitary engineering literature, so they will not be discussed in detail here. A few general comments and some specific references for more detailed design information are presented.

There are two general schools of thought on trickling filter design: one based on organic loading and the other based on hydraulic loading. Design models based on organic loading have been proposed by several design textbooks[2,8,9] and regulatory agencies,[10] and have been used by several manufacturers of trickling filter media[11,12] for many years. Other well-known authorities in the field,[13-17] as well as several media manufacturers,[18,19] have proposed hydraulic loading models, and one authority has used both hydraulic and organic loading models.[20] Organic loading models predict BOD removal efficiency as a function of applied organic load expressed as pounds of 5-day BOD/day/1,000 ft^3 of media volume or other equivalent units of mass per unit time per unit volume. Hydraulic loading models express BOD removal efficiency as a function of wastewater retention time, which in turn is usually expressed as a function of the hydraulic loading on the top filter area (gal/min/ft^2) and the filter depth (14 to 30 ft for synthetic media).

Both hydraulic and organic loading models can be useful, depending upon the nature of the wastewater and required degree of treatment. The part of the BOD content of wastewater associated with suspended and colloidal matter is removed quickly by adsorption and flocculation. Longer wastewater contact times are required to remove a large fraction of the soluble BOD constituents. This implies that organic loading models will be more accurate for applications at high loadings and low degrees of treatment, and that hydraulic loading models will be more accurate for applications at low loadings and high degrees of treatment for wastewater with significant quantities of both soluble and suspended BOD. (This would apply to domestic waste which has about half soluble and half suspended BOD after primary treatment.) Similarly, BOD removal on wastewater with high suspended BOD content would be best determined with organic loading models, and for those with high soluble BOD content, BOD removal would be best determined with hydraulic loading models.

A summary of design criteria for domestic waste treatment with organic loading models is presented in the *Handbook of Trickling Filter Design.*[2] Design criteria for treatment of various industrial wastes using organic loading criteria have been presented by Askew and Chipperfield.[21-23] Several nomographs for designing trickling filters have been developed using both hydraulic and organic loading criteria and including the effect of wastewater temperature.[2,15]

The design model most often mentioned in the literature is an organic loading model developed from field operating data by the National Research Council in 1946 and called the NRC formula.[24] When both this model and a hydraulic loading model were recently compared to the original NRC data, they were judged to be statistically unacceptable in predicting trickling filter performance on domestic waste.[25] This same report also stated that effluent recirculation did not show any significant improvement in performance or any significant improvement in predicting performance. While this may not be true for all data collected since 1946, it does point out a problem in reconciling field performance with design models and calls for further modeling work using field rather than laboratory test data.

CAPITAL AND OPERATING COSTS

Capital cost for trickling filter systems has been increasing during the last several years. Part of this increase has been due to the increased cost of plastic materials often used as the trickling filter media. In the early 1970s, plastic media costs were about $1.50/ft^3. This has increased to $3.00/ft^3 and above in 1975. The total installed cost for a trickling filter system varies from $6.00 to $10.00/ft^3 which includes the tower and effluent recirculation system. The upper range of these costs apply to multistage trickling filter systems, which are usually necessary to achieve high degrees of BOD removal and nitrification.

Operating costs for the trickling filter system

consist chiefly of power consumption for waste-water pumping and effluent recirculation. The amount of power consumption is influenced significantly by the amount of effluent recirculation, which can be anywhere from 50 to 300% of the wastewater flow. This becomes especially significant in multistage systems where several recirculation systems are used. A recent cost study using several trickling filter design models has shown that the improvement in treatment efficiency from effluent recirculation will not be justified in many cases because of the higher capital and operating costs of the effluent recirculation system.[26]

REFERENCES

1. Grieves, C. R., Dynamic and Steady State Models for the Rotating Biological Disc Reactor, Ph.D. thesis, Clemson University, Clemson, S.C., 1972.
2. *Handbook of Trickling Filter Design,* Public Works Corp., Ridgewood, N.J., 1970.
3. Chase, S. E., Trickling filters – past, present, and future, *Sewage Works J.,* 17, 929, 1945.
4. EPA Technology Transfer: The bridge between research and use, EPA Rep., March 1, 1973.
5. Hartman, H., *Investigation of the Biological Clarification of Wastewater Using Immersion Drip Filters,* Vol. 9, Suttgarter Berichte zur Siedlungswasserirtschaft, R. Oldenbourg, Munich, 1960.
6. Hang, R. and McCarty, P., Nitrification with submerged filters, *J. Water Pollut. Control Fed.,* 44(11), 2086, 1972.
7. Jeris, J., Owens, R., and Flood, F., Biological Fluidized Bed Technology, paper presented at the 48th Annu. Water Pollut. Control Fed., Oct. 5–10, 1975, Miami Beach, Fla.
8. Fair, G., Geyer, J., and Okun, D., Water and Wastewater Engineering, Vol. 2, John Wiley & Sons, New York, 1968, 34.
9. Sewage treatment plant design, WPCF Manual of Practice No. 8, ASCE Manual of Eng. Practice. No. 36, New York, 1959, 151.
10. Recommended Standards for Sewage Works (Ten States Standards), Great Lakes – Upper Mississippi River Board of State Sanitary Engineers, Health Evaluation Service, Albany, N.Y., 1968, 69.
11. Ethyl Corp., Brochure ECD-971, Baton Rouge, La.
12. Del-Pak Corp. Design Catalog, Technol. Rep. No. 106-5-72, Corvallis, Ore.
13. Eckenfelder, W. W., Trickling filter design and performance, *J. Sanit. Eng.,* 87, SA6, 1961.
14. Swilley, E. and Atkinson, B., A Mathematical Model of the Trickling Filter, Proc. 18th Purdue Ind. Waste Conf., May 6–8, 1963, W. Lafayette, Ind., 706.
15. Mehta, D., Davis, H., and Kingsbury, R., A new design tool for plastic media trickling filters, B.F. Goodrich Co. Publ. IPC-1068-6, reprinted from *Ind. Water Eng.*
16. McDermott, J. H., Influence of Media Surface Area upon the Performance of an Experimental Trickling Filter, Thesis, Purdue University, W. Lafayette, Ind., 1957.
17. Keefer, C. and Meisel, J., *Water Sewage Works,* 99, 277, 1952.
18. B.F. Goodrich Design Catalog, p. 1–1, Feb., 1972.
19. Koch Engineering Co., Bull. TRF-1, New York.
20. Germain, J. E., Economical treatment of domestic waste by plastic medium trickling filters, *Water Pollut. Control Fed.,* 38(2), 192, 1966.
21. Chipperfield, P. N., Performance of plastic filter media in industrial and domestic waste treatment, *J. Water Pollut. Control Fed.,* 39(11), 1860, 1967.
22. Chipperfield, P. N., Askew, M., and Benton, J., Multi-stage Plastic Filter Media Proving Effective in Treating Biodegradable Industrial Wastes, Proc. 25th Purdue Ind. Waste Conf., May 5–7, 1970, W. Lafayette, Ind; *Ind. Water Eng.,* August 1970.
23. Askew, M. W., Biofiltration technology, *Effluent Water Treat. J.,* Oct. and Nov. 1969.
24. Subcommittee on Sewage Treatment, Committee on Sanitary Engineering, National Research Council, Sewage treatment at military institutions, *Sewage Works J.,* 18, 789, 1946.
25. Schroeder, E. and Tchobauoglous, G., Another look at the NRC formula, *Water Sewage Works,* 122, 58, 1975.
26. Lee, C. and Takamatsa, T., Cost of trickling filter recirculation, *Water Sewage Works,* Part 1, 122, 57, 1975; Part 2, 122, 64, 1975.

Chapter 2

THE ROTATING BIOLOGICAL CONTACTOR HISTORY AND PROCESS FUNDAMENTALS

PROCESS DESCRIPTION

The rotating biological contactor process is a secondary biological wastewater treatment system. It consists of large-diameter plastic media, which is mounted on a horizontal shaft and placed in a concrete tank as shown in Figure 2-1. The contactor is slowly rotated while approximately 40% of the surface area is submerged in the wastewater. Immediately after start-up, organisms naturally present in the wastewater begin to adhere to the rotating surfaces and multiply until, in about 1 week, the entire surface area is covered with an approximately 1- to 4-mm thick layer of biomass (see Figure 2-2). The attached biomass contains approximately 50,000 to 100,000 mg/l suspended solids. If the biomass were detached and dispersed in the mixed liquor, the resulting mixed-liquor suspended-solids concentration would be 10,000 to 20,000 mg/l. This large microbial population permits high degrees of treatment to be achieved for relatively short wastewater retention times.

In rotation, the contactor carries a film of wastewater into the air, which trickles down the surfaces and absorbs oxygen from the air. Organisms in the biomass then remove both dissolved oxygen and organic materials from this film of wastewater. Further removal of organic materials and consumption of dissolved oxygen occurs as the surfaces continue rotation through the bulk of the wastewater in the tank. Unused dissolved oxygen in the wastewater film is mixed with the contents of the mixed liquor, which maintains a mixed-liquor dissolved-oxygen concentration.

Shearing forces exerted on the biomass as it passes through the wastewater cause excess biomass to be stripped from the media into the mixed liquor. This prevents clogging of the media surfaces and maintains a constant microorganism population on the media. The mixing action of the rotating media keeps the stripped solids in suspension until the flow of treated wastewater carries them out of the process for separation and disposal. Operating in this manner, the rotating media serves the following functions:

1. Providing surface area for the development of a large, fixed biological culture.

2. Providing vigorous contact of the biological growth with the wastewater.

3. Efficiently aerating the wastewater.

4. Providing a positive means of continuously stripping excess biomass.

5. Agitating the mixed liquor to keep sloughed solids in suspension and to thoroughly mix each stage of treatment.

The nature of the attached biological growth is much different from that which develops in trickling filter systems. Figure 2-3 is an attempt to depict this difference. In the trickling filter, the growth is uniform in thickness and often has a gelatinous appearance, which is why the growth is usually referred to as a slime. The only way soluble substrate and dissolved oxygen can be supplied to the portion of the biomass close to the support medium is through diffusion. Because the outer portion of the biomass quickly consumes most of the available oxygen, the inner portion is often anaerobic. This anaerobic activity is thought to be the principal mechanism for the periodic sloughing or dislodging of large amounts of the growth. With oxygen and substrate limited to just the outer portion of the growth, only that fraction of the total biomass is actively treating the wastewater.

In contrast, Figure 2-3 shows that the biomass of the rotating contactor is shaggy with many macroscopic filaments, which project outward into the adjacent film of wastewater. This provides an active biological surface area much larger than just the surface area of the support medium. It also enables substrate and dissolved oxygen to reach a greater portion of the biomass and render it aerobically active.

The shagginess is not due to the presence of any special species of organisms but to the action of rotation. The continual drag from being rotated through the wastewater, and the draining of entrained wastewater when rotated up into the air, cause the growth to form the elongated macroscopic filaments. This shaggy structure is most apparent in the initial stages of the media, where

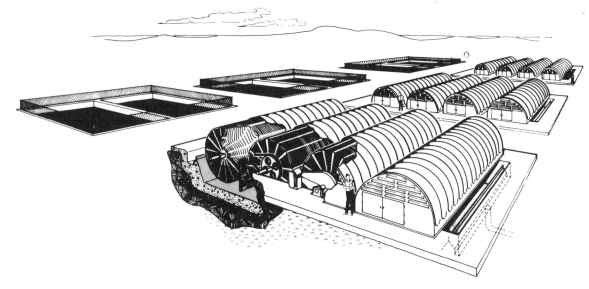

FIGURE 2-1.   RBC treatment plant.

FIGURE 2-2.   Attached biomass.

the BOD concentrations are higher and the growth is thicker, and gradually decreases in the later stages, where BOD concentrations are lower and growths are thinner. Unlike the trickling filter, the principal mechanism of displacing excess biological growth with the rotating contactor is through hydraulic shear. The shaggy growth will increase in length and thickness until it can no longer withstand the shear exerted by rotation. The growth that is stripped off is in the form of relatively large aggregates of dense biomass, which settle rapidly in a final clarifier.

Rotation of the media also provides turbulence at the interface between biomass and wastewater,

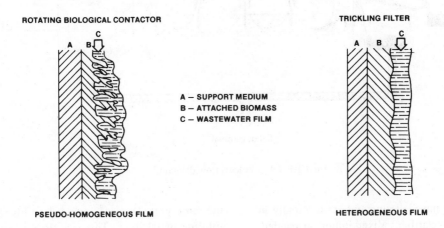

FIGURE 2-3.   Fixed-film biological reactors.

so that dissolved oxygen and soluble substrate are available to the inner portion of the biomass through the mechanism of mixing as well as that of diffusion. At a peripheral media velocity of 1.0 ft/sec, the Reynolds number for flow through the trapezoid-shaped concentric corrugations is well over 7,000 at the outer perimeter, and over 5,000 at the centroid of the surface area (a point about 2/3 of the radius from the shaft center). This puts the flow well beyond the threshold of laminar-to-turbulent flow (Reynolds number 2,100) and into the regime of near-turbulent flow.

PROCESS OPERATION

Figure 2-4 shows a process flow diagram for a treatment plant incorporating the rotating contactor process. Raw wastewater flows first through primary treatment for removal of large objects and floatable and settleable materials. Primary effluent then flows to the multistage fixed-film process, where aerobic cultures of microorganism remove both dissolved and suspended organic matter from the wastewater. Part of the organic matter is oxidized to carbon dioxide and water, part is synthesized into additional biomass, and part is stored in the biomass for oxidation and synthesis at a later time. Simultaneously, as additional biomass is being generated, excess biomass is being

stripped from the rotating media by the shearing forces exerted by the wastewater.

Each shaft of media operates as a completely mixed, fixed-film biological reactor, in which the rate of biological growth and the rate of stripping excess biomass are at a dynamic equilibrium. Treated wastewater and stripped biomass pass through each subsequent stage of media. As wastewater passes from stage to stage, it undergoes a progressively increasing degree of treatment by specific biological cultures in each stage, which are adapted to the changing wastewater. Initial stages of media, which receive the highest concentration of organic matter, develop cultures of filamentous and nonfilamentous bacteria. As the concentration of organic matter decreases in subsequent stages, higher life forms including nitrifying bacteria begin to appear, along with various types of protozoans, rotifers, and other predators.

Excess biomass and treated wastewater leaving the last stage of media pass to a secondary clarifier, where the solids are separated for disposal. Clarified effluent passes on for disinfection or further treatment. Settled solids thicken to a concentration of 3 to 4% solids in the secondary clarifier. When secondary solids are recycled to the primary clarifier (when treating domestic wastewater), a combined sludge of 4 to 6% solids is normally produced for disposal.

Stripped biological solids retain their high

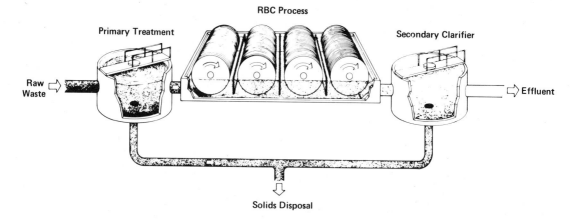

FIGURE 2-4.   Process flow diagram.

density in the mixed liquor and settle rapidly in the secondary clarifier. Mixed-liquor suspended-solids concentrations are approximately one half of the influent BOD concentration and, depending on the degree of treatment, will vary from 50 to 200 mg/l in the treatment of domestic wastewater.

Wastewater flows through the process just once, with no need for recycling of effluent. Because the attached biomass is continuously growing, there is also no need for recycling of sludge. The absence of any recycle requirements makes operation of the process very simple. Use of standard electric motor and gear reducer drives with chain and sprocket final drive and low contactor rotational speeds result in very low maintenance requirements. Because of the operational simplicity, a relatively low-skilled plant operator can successfully operate a rotating contactor plant. This is very important for small communities, which cannot afford or obtain skilled operators for wastewater treatment plants.

HISTORY

The rotating contactor for wastewater treatment was first conceived in Germany by Weigand in 1900. His patent for the contactor[1] describes a cylinder consisting of wooden slats. However, no units were constructed until the 1930's when Bach[2] and Imhoff[3] tested them as substitutes for the Emscher filter. These units experienced severe problems with clogging of the slats and were not investigated further.

In the U.S., Allen reported the invention of the "biological wheel" by Maltby in 1929.[4] It consisted of a series of rotating paddle wheels. In the same year, Doman[5] reported on his testing of rotating metal discs. This was the first time that discs were investigated as the contact media, but the results were not encouraging, and no further work was done in the U.S.

Nothing further developed in Europe until the late 1950s, when first Hartman[6] and then Pöpel[7] at Stuttgart University conducted extensive testing using plastic discs, 1.0 m in diameter. About the same time, expanded polystyrene came into use as an inexpensive construction material. The process development work of Hartman and Pöpel, together with this new material of construction, resulted in a new commercial wastewater treatment process. For more details on the history of the rotating biological contactor, refer to Hartman[6] and to Grieves.[8]

The J. Conrad Stengelin Co. in Tuttlingen, West Germany, began manufacturing expanded-polystyrene discs 2 m and 3 m in diameter for use in wastewater treatment plants in 1957. The first commercial installation went into operation in 1960, and use of the process spread quickly throughout Europe because of its simplicity and low power consumption. At the present time there are almost 1,000 installations, located primarily in West Germany, Switzerland, and France, with a few installations in Italy, Austria, Great Britain, and the Scandinavian countries. The majority of these installation are for populations of less than 1,000, with some as small as a single residence. There are only a few installations of more than 10,000 population equivalent. Although offering greater simplicity in operation and much lower power requirements, the process has been restricted to relatively small installations because of its high construction cost relative to activated sludge systems.

Development work on the rotating-disc process began in the U.S. at Allis-Chalmers in the mid 1960s. This was done without knowledge of the previous work and was an outgrowth of other work on testing of rotating discs for chemical processing applications. After learning of the European activities, Allis-Chalmers reached a licensing agreement with the German manufacturer for manufacturing and sales distribution in the U.S. The treatment process was marketed under the trade name BIO-DISC in both the U.S. and Europe. The first commercial installation in the U.S. went into operation at a small cheese factory in 1969.

Activities continued at Allis-Chalmers until 1970 with only limited commercial success. At that time the business was purchased from Allis-Chalmers by Autotrol Corporation. Use of the expanded-polystyrene disc construction continued with some commercial success, but, as in Europe, it was limited to small domestic and industrial waste applications. In 1972, Autotrol announced the development of new rotating contactor media constructed from corrugated sheets of polyethylene. This increased the surface area density from 16 ft^2/ft^3 for the polystyrene disc to 37 ft^2/ft^3 for the new corrugated media. Increasing the contactor diameter to 12 ft and the length to 25 ft provided as much as 100,000 ft^2 of surface area available on one contactor. The corrugated-media construction is now being marketed under the trade name BIO-SURF process.

Since this development, commercial use of the process has increased steadily. There are now more than 80 installations in the U.S. and Canada. Fifteen have more than 1 mgd capacity. Two plants under construction in 1975 will be treating 30 mgd of pulp and paper waste and 54 mgd of municipal waste.

In recent years, other manufacturers in the U.S. and Canada have begun to produce expanded-polystyrene discs, which are being marketed as the BIO-DISC process and as the Rotating Biological Surface process or RBS process.

During the initial research work at Allis-Chalmers, the process was referred to as the rotating biological contactor or RBC. Since this name is not associated with any particular manufacturer, it will be used as a generic term for the process in subsequent chapters.

COMPARISONS WITH OTHER SECONDARY TREATMENT PROCESSES

The rotating contactor process is similar in function to the trickling filter process in that both operate as fixed-film biological reactors. Passing wastewater over the media, as in a trickling filter, results in a near-laminar flow of wastewater down through the media. This affects the trickling filter process operation in several ways.

Short contact time — Wastewater retention time in a trickling filter is relatively short, due to the small wastewater inventory in the media at any given time. This results in limited degrees of treatment especially for soluble BOD removal.

Poor wastewater contact — Contact between the biological growth and the wastewater in a trickling filter is not intense, resulting in little penetration of organic matter and dissolved oxygen into the biological growth.

Ineffective sloughing — Excess growth is not effectively removed in a trickling filter, so that clogging can occur. Sloughing is often accomplished through the development of anaerobic conditions at the interface between the biological growth and the filter media.

In a rotating contactor, the biomass is passed through the wastewater rather than the wastewater passed over the biomass. This key difference results in a number of improvements.

Controlled sloughing — Clogging of the media is prevented, because the shearing forces continuously and uniformly strip excess growth from the media.

No nuisances — Continuous wetting of the entire biomass prevents development of "filter flies," often associated with trickling filter operation.

Efficient aeration — Aeration with the rotating media is a very positive means of supplying dissolved oxygen to all biomass-covered surfaces.

Controlled contact and aeration — Both the intensity of contact between the biomass and the wastewater and the aeration rate can be easily controlled by designing for an appropriate rotational speed.

High degrees of treatment — Wastewater retention time is also controlled by selecting an appropriate tank size. Thus, very high degrees of treatment can be attained. It is unnecessary to recycle effluent in order to achieve minimum wetting rates, enrich influent dissolved-oxygen concentration, or aid in sloughing, as is required by the trickling filter. This allows the rotating contactor process to take advantage of the benefits of staged operation, which would otherwise be destroyed by effluent recycle.

The rotating contactor process is similar to the activated sludge process in that it has some suspended culture in its mixed liquor and can produce high degrees of treatment and a sparkling clear effluent. However, the major difference is that more than 95% of the biological solids in this fixed-film system are attached to the media, which leads to the following major process differences.

Process stability — The activated sludge process depends on sludge recycling for satisfactory operation. A hydraulic surge can result in significant loss of sludge over the secondary clarifier weir and cause a prolonged upset in the process operation. Organic shock loads can also cause loss of sludge from the secondary clarifier through sludge bulking. The rotating contactor process is not upset by variations in hydraulic or organic loading, because the majority of the active organisms are attached to the media.

Flexibility — Two major problems are encountered in operating small activated sludge treatment plants: start-up at flows much lower than design flow, and operation during periods of little or no flow. Operation of a rotating contactor plant at low initial flows or during periods of very low flow will yield effluents of higher quality than at design flow. During no-flow conditions, effluent or sludge treatment supernatant can be recycled to provide some organic matter to maintain biological activity. Also, the process lends itself better to upgrading existing treatment facilities because of its modular construction, low hydraulic head loss,

and shallow excavation. Shaft assemblies can be integrated with existing primary and secondary facilities to upgrade treatment levels and minimize additional construction and land requirements.

Maintenance and power consumption — The low maintenance and low power consumption of the rotating contactor system are two of its most attractive features when compared to the activated sludge process.

Ease of nitrification — Many state regulatory agencies are requiring treatment plants to be built to achieve various degrees of nitrification as well as BOD and suspended-solids removal. To achieve this with the activated sludge process usually requires that an expensive two-stage process be constructed, with separate aeration, settling, and sludge recycle systems. The rotating contactor process has demonstrated that it can achieve any desired degree of nitrification in a single treatment step, with only a single settling tank and without sludge recycle.

Sludge characteristics — The high density and low concentration of mixed-liquor solids in the rotating contactor allows secondary clarifiers to be designed for relatively high overflow rates and without concern for problems of hindered settling and sludge compression, as is the case with the activated sludge process. Also, secondary clarifiers need not be designed to accommodate the high sludge recycle ratio required for the activated sludge process. Unlike the activated sludge process, the rotating contactor process can be designed for any degree of treatment, and the secondary sludge will settle well. Because the sludge will thicken to high concentrations in the secondary clarifier, sludge-thickening requirements can be reduced or eliminated for many applications.

REFERENCES

1. **Weigand, Ph.,** Verfahren zur Biologischen Reinigung von Abwassern, German Patent No. 135755, 1900.
2. **Bach, H.,** *Waste Water Clarification,* Vol.2, R. Oldenbourg, Munich, 1934, pp. 147—150.
3. **Imhoff, K.,** Bio-Immersion Filters, Technical Community Papers 29, 1926, #13.
4. **Allen, K.,** *The Biologic Wheel,* Sewage Disposal Bulletin, City of New York, No. 14, 1929; *Sewage Works J.* 1 (5), 560, 1929.
5. **Doman, J.,** Results of operation of experimental contact filter with partially submerged rotating plates, *Sewage Works J.,* 1 (5), 555, 1929.
6. **Hartman, H.,** *Investigation of the Biological Clarification of Wastewater Using Immersion Drip Filters,* Vol. 9, Stuttgarter Berichte zur Siedlungswasserwirtschaft, R. Oldenbourg, Munich, 1960.
7. **Pöpel, F.,** *Estimating, Construction and Output of Immersion Drip Filter Plants,* Vol. 11, Stuttgarter Berichte zur Siedlungswasserwirtschaft, R. Oldenbourg, Munich, 1964.
8. **Grieves, C.,** Dynamic and Steady State Models for the Rotating Biological Disc Reactor, Ph.D. thesis, Clemson University, Clemson, South Carolina, August, 1972.

Chapter 3

PROCESS DEVELOPMENT

INTRODUCTION

Pilot plant testing of the rotating contactor process on domestic wastewater has been conducted in the U.S. since 1969. The following discussion reviews the major pilot plant tests and the process development they produced.

INITIAL PROCESS DEVELOPMENT STUDIES

In December 1969, the pilot plant shown in Figure 3-1 was started up at Pewaukee, Wisconsin, and testing continued through October 1971, under sponsorship of EPA Research Contract 14-12-810.[1] The pilot plant consists of a wet well and rotating bucket feed system, a multistate rotating disc treatment system, and a secondary clarifier with a rotating sludge scoop, all incorporated into a single tank of semicircular shape. This prototype package plant is intended to operate in conjunction with primary treatment and sludge disposal facilities.

In operation, primary treated wastewater flows into the wet well, where a series of buckets, attached to the same shaft that rotates the discs, deposits it via a trough into the first stage of discs. An overflow connection maintains a constant level in the wet well, and therefore a constant feed rate. The feed rate is varied by changing the number of feed buckets.

The disc section contains approximately 90 discs of 1.75 m diameter, which are divided into several equal-size stages. Each disc has approximately 51 ft² of surface area available for biological growth. The stages are divided by bulkheads along the length of the tank and connected by a trough and submerged orifices so that they operate in series. The discs, which are molded from expanded-polystyrene beads, are about 0.5 in. thick and spaced on 1.0-in. centers.

Power is supplied to the discs by a $\frac{1}{6}$ hp electric motor, which is mounted on a structure between the disc section and the secondary clarifier. The power is transmitted to the main shaft by the friction of a chain passing over a circular metal strip of the same diameter as the discs, which is mounted on the last set of radial arms (see Figure 3-2).

Mixed liquor passes from the last stage of discs into the clarifier through an opening in the bulkhead separating them. To minimize the effects of turbulence, the opening is located in the corner of the clarifier, and a pipe directs the flow against the bulkhead beneath the water level.

FIGURE 3-1.   RBC pilot plant.

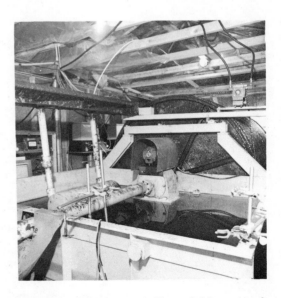

FIGURE 3-2.   Secondary clarifier and drive system for pilot plant.

Settled sludge is removed by a rotating scoop and reservoir system as shown in Figure 3-2. The scoop system is driven independently at 2 to 6 rph to allow adjustment for sludge removal requirements. As the empty reservoir passes into the clarifier, water pushes collected sludge up through the scoop and hollow connecting arms to fill the reservoir. As the reservoir leaves the water, it empties the sludge through hollow arms into the hollow drive shaft. From there, it flows out of the system for treatment and disposal. Clarifier overflow passes over a weir to an outlet located in the opposite corner from the inlet.

The clarifier has an operating volume of 456 gal and an overflow area of 32 ft.^{2} The overall size of the package unit is approximately 16 ft long, 6.5 ft wide, and 6.5 ft high. It has a design flow of 10,000 gpd.

The pilot plant was installed in a garage structure to protect it from wind, heavy precipitation, and freezing temperatures, and was located within the grounds of the wastewater treatment plant of the village of Pewaukee, Wisconsin. A portion of the effluent from the primary clarifier was used as the wastewater source for the testing. The primary effluent had the characteristics shown in Table 3-1.

Effect of Rotational Speed on Treatment Capacity

The first phase of the test program was conducted with the discs in the pilot plant divided into two equal-size stages in series. The principal variable investigated was the effect of rotational

disc velocity on treatment capacity. Disc speeds of 4.6, 3.2, and 2.0 rpm were tested, which, on the 5.74-ft-diameter discs, resulted in peripheral velocities of approximately 83, 58, and 36 ft/min, respectively. BOD removal as a function of hydraulic loading (measured as gallons per day of wastewater flow per square foot of growth-covered surface area) is shown in Figure 3-3 for the three speeds tested. The two higher speeds gave very similar BOD removals over a wide range of hydraulic loadings, whereas the lowest speed resulted in lower degrees of treatment over the same range.

Rotational disc velocity affects treatment in several ways. It increases the intensity of contact between the biomass and the wastewater; it increases the rate of aeration; and it more thoroughly mixes the contents of each stage of treatment. These factors appear to affect treat-

TABLE 3-1

Wastewater Characteristics after Primary Treatment

	Average (mg/l)	Range (mg/l)
BOD	150	100–250
Suspended solids	120	50–250
COD	350	150–500
Ammonia nitrogen	18	10–25
Kjeldahl nitrogen	29	15–40
Phosphorus	10	5–12

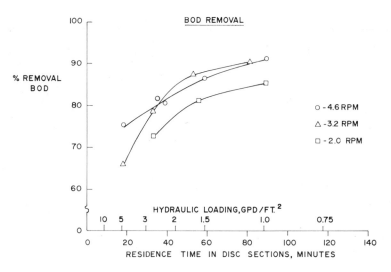

FIGURE 3-3. BOD removal in two-stage pilot plant.

ment up to a certain point, beyond which further increases in rotational speed have no further effect on treatment capacity.

Figure 3-4 shows removal of ammonia nitrogen for the same three disc speeds. Here, the two higher speeds gave virtually identical performance over the full range of loadings, and the lowest speed showed significantly lower ammonia removal. Based on the results of these tests, it is concluded that the peripheral velocity associated with 3.2 rpm, i.e., approximately 60 ft/min, is a basic design criterion for domestic wastewater treatment.

Effect of Two-stage and Four-stage Operation on Treatment Capacity

For the second phase of the studies,[2] the same test unit was divided into four stages with the insertion of two additional bulkheads in the tank. Figure 3-5 shows BOD removal as a function of hydraulic loading for both two- and four-stage operation, and it is apparent that four-stage operation makes more efficient use of the surface area. For example, at 3.0 gpd/ft², the two-stage system removed approximately 75% of the BOD, whereas the four-stage system removed almost 85%. Conversely, for 85% BOD removal, a two-stage system would have to be operated at a hydraulic loading of approximately 1.8 gpd/ft², whereas a four-stage system could be operated at almost 3.0 gpd/ft². Figure 3-6 shows a similar effect of staging on suspended-solids removal.

The primary reason for the improved performance of four-stage over two-stage operation is that operation in a larger number of stages in series improves the residence time distribution, or more closely approximates a "plug flow." Because BOD removal rate is concentration-dependent, this increases the overall BOD removal.

A comparison of the relative efficiency of two-stage and four-stage operation (based on Figure 3-5) is shown in Figure 3-7. When designing for very high degrees of treatment, staging is very important for maximizing the effectiveness of the rotating surface area. For low degrees of treatment, however, staging is relatively unimportant.

The rotating disc process appears to be especially suited to obtaining nitrification of domestic wastewater. Because the process is constructed in a series of stages, the fixed cultures that develop in each successive stage become adapted to treating the wastewater as it undergoes a progressively increasing degree of treatment. The majority of the carbonaceous BOD is satisfied in the initial stages of discs. This allows nitrifying bacteria to predominate in subsequent stages for achievement of a high degree of ammonia oxidation. This occurs without intermediate clarification or sludge recycle.

Figure 3-8 shows ammonia nitrogen removal for two-stage and four-stage operation. Nitrification is essentially identical for either mode of operation and indicates that high degrees of nitrification can be achieved in as few as two stages of treatment.

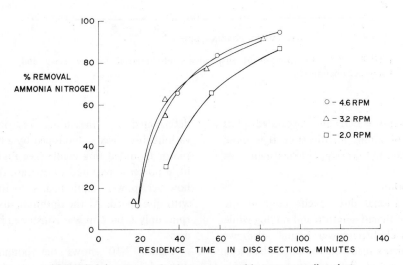

FIGURE 3-4.   Ammonia nitrogen removal in two-stage pilot plant.

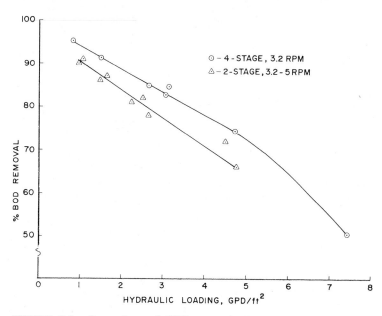

FIGURE 3-5. Comparison of BOD removals for two-stage and four-stage operation.

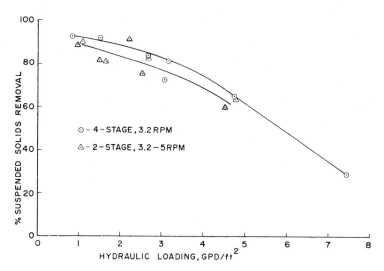

FIGURE 3-6. Comparison of suspended-solids removal for two-stage and four-stage operation.

However, to maximize BOD and suspended-solids removals while obtaining nitrification, it is more efficient to operate in four stages of treatment.

Power Consumption

The low rotational disc speeds used in the process achieve efficient aeration and mixing while consuming relatively little power. Figure 3-9 shows power measurements made on the pilot plant as a function of disc speed. Power requirements were determined by measuring reaction torque at various disc speeds, developed by a hydraulic drive motor mounted in a cradle (see Figure 3-2). Very little power is required to operate the pilot plant; however, power requirements do increase rapidly with disc speed. At the optimum disc speed of 3.2 rpm, only 0.06 bhp was consumed by the rotating discs.

Figure 3-10 shows the pounds of oxygen demand removed per horse-power-hour of energy

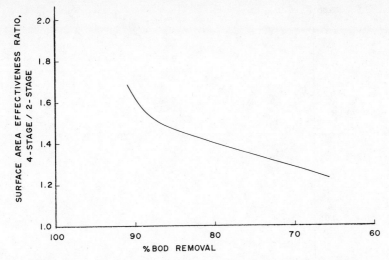

FIGURE 3-7.   Relative effectiveness of four-stage and two-stage operation.

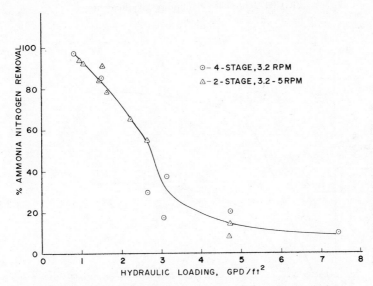

FIGURE 3-8.   Nitrification in two-stage and four-stage operation.

input as a function of BOD removal. It is shown for both 5-day BOD and ultimate oxygen demand.* At 85% BOD removal, approximately 10 lb of BOD5 are removed per hp-hr. As the degree of treatment increases, power consumption also increases until at 95% BOD removal and virtually complete nitrification, 3 lb of BOD_5 and 6 lb of UOD are removed per hp-hr.

The previous correlations have shown BOD and

suspended solids removal and nitrification at 3.2 and 4.6 rpm to be superior to that at 2.0 rpm. In spite of this, Figures 3-11 and 3-12 show that a speed of 2.0 rpm makes more efficient use of power for all the treatment levels and effluent concentrations obtained. However, considering that a much smaller treatment plant is necessary for a given application, the increased power consumption at higher disc speeds seems justified.

*Ultimate oxygen demand is defined by

UOD = 1.5 × BOD5 + 4.5 × NH3 − N,

where NH3 − N is the ammonia content in mg/l as nitrogen.

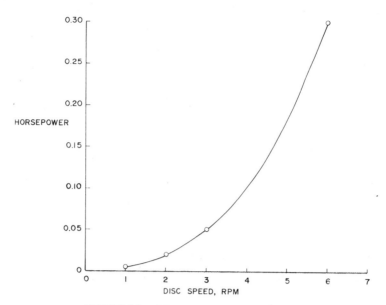

FIGURE 3-9. Pilot plant power requirements.

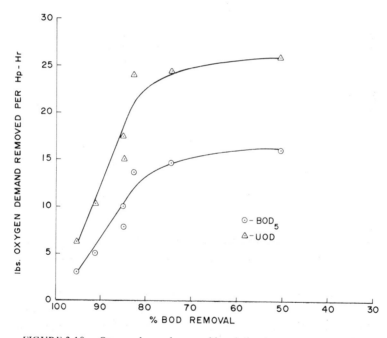

FIGURE 3-10. Oxygen demand removal in relation to power consumption.

Oxygen Demand Removal Kinetics

Throughout this discussion, hydraulic loading has been used as the primary criterion for the rotating contactor process. The reason for this is that the process is approximately first order with respect to BOD removal. This means that, at a given hydraulic loading, there will be a given percentage of reduction of BOD, independent of the influent BOD concentration.

One way of showing that the process is first order with respect to BOD removal is to plot percent BOD remaining as a function of retention time on semilog paper. A resulting straight line indicates first order behavior, and the slope of the line is the reaction velocity constant or "K" factor. The solid line in Figure 3-13 shows that the process is first order in removal of BOD for approximately the first 85%. Beyond that point,

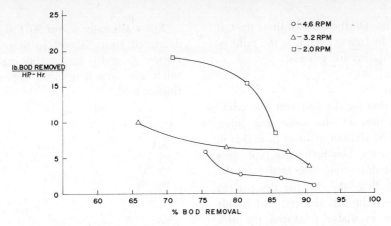

FIGURE 3-11. Comparison of power requirements for different disc speeds.

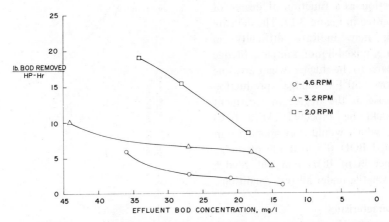

FIGURE 3-12. Comparison of power requirements for different disc speeds.

the removal continues to be first order, but with a lower velocity constant. Other BOD analyses were performed with allythiourea added to the dilution water to suppress nitrification during the 5-day incubation period. This yielded a measure of carbonaceous BOD alone. The broken line in Figure 3-13 shows that the process is first order in carbonaceous BOD removal up to 90%. Beyond that point, the removal continues to be first order, but again with a lower velocity.

Change in the velocity constant at approximately 30-min retention time is due to the change in the microbial population from one that chiefly oxidizes carbonaceous BOD to one that oxidizes nitrogenous BOD. Five-day BOD removals of 84 and 91% then correspond to carbonaceous BOD removals of 90 and 96%, respectively.

This change in the population appears to be a function of BOD concentration. Figure 3-14 shows

the degree of nitrification as a function of effluent BOD concentration for two-stage and four-stage operation. Ammonia nitrogen removal is relatively constant for effluent BOD values above 40 mg/l and is due primarily to cell synthesis. Below 40 mg/l for two-stage operation and below 30 mg/l for four-stage operation, nitrifying bacteria begin to predominate, and ammonia oxidation proceeds rapidly until, at an effluent BOD of 5 to 10 mg/l, nitrification is virtually complete.

Plotting the percentage of ammonia nitrogen remaining as a function of retention time on semilog paper, as in Figure 3-15, shows that once ammonia nitrogen oxidation begins, it proceeds in a first-order mechanism up to at least 97% removal.

Plotting ultimate oxygen demand* in Figure 3-16 in the same fashion as Figures 3-13 and 3-15 shows that UOD removal is first order up to at

*Ultimate oxygen demand is defined by
$$UOD = 1.5 \times BOD_5 + 4.5 \times NH_3 - N$$
where $NH_3 - N$ is the ammonia content in mg/l as nitrogen.

21

least 93% reduction. On the basis of these tests, it is concluded that hydraulic loading is the primary criterion for the rotating disc process.

Sludge Production

Sludge production by the test unit was calculated by subtraction of the suspended solids concentration in the clarifier effluent from that in the last stage of discs. This difference represents the amount of sludge solids that settled in the clarifier. Dividing this value by the decrease in BOD concentration through the test unit yields sludge production as sludge produced per unit BOD removed.

Sludge production as a function of degree of treatment is presented in Figure 3-17. The data are scattered, which may indicate difficulty in obtaining accurate mixed-liquor samples. Sludge production appears to be higher when treating wastewater below 50°F. Sludge production appears to decrease as the degree of treatment increases, as would be expected. At a BOD removal of 86%, which would correspond to an effluent of 20 mg/l BOD, 0.5 to 0.6 lb of sludge were produced per lb of BOD removed. Sludge solids were 80% volatile under all test conditions.

Mixed Liquor Characteristics

The parameters of pH, dissolved-oxygen concentration, and suspended solids for the mixed liquor of the two stages of discs are summarized in Table 3-2.

For wastewater above 50°F, suspended matter decreased from Stage 1 to Stage 2 to a concentration generally under 100 mg/l, mixed-liquor solids are very dense and appear to settle and thicken well.

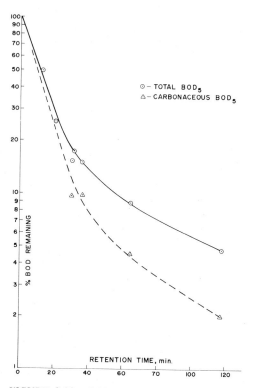

FIGURE 3-13. BOD removal kinetics as a function of retention time.

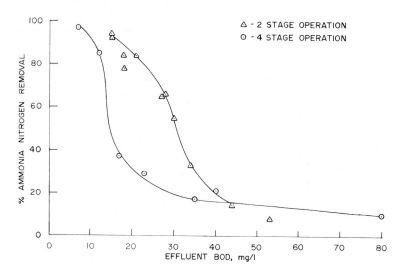

FIGURE 3-14. Effect of BOD concentration on nitrification.

FIGURE 3-25.   Pilot plants at high hydraulic loading.

FIGURE 3-26.   Pilot plants at low hydraulic loading.

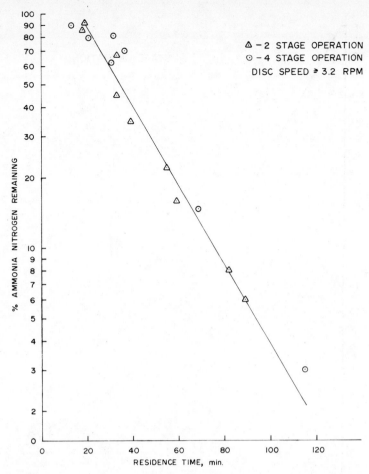

FIGURE 3-15.   Ammonia nitrogen removal kinetics as a function of retention time.

For wastewater temperatures below 50°F, suspended matter generally increased from Stage 1 to Stage 2 to concentrations ranging from 130 to 250 mg/l; this accounts for the higher calculated sludge production rate at lower wastewater temperatures.

Dissolved-oxygen concentration in Stage 1 varied from 1.0 to 3.8 mg/l, depending on hydraulic loading and disc speed. Stage 2 dissolved-oxygen concentrations were 1 to 2 mg/l higher than Stage 1. Less than 1.0 mg/l of dissolved oxygen was consumed in the secondary clarifier, so that effluent dissolved-oxygen concentrations were always several milligrams per liter. At wastewater temperatures below 50°F, dissolved-oxygen concentrations in the effluent ranged between 7 and 10 mg/l, which indicates that mixed-liquor concentrations were also quite high.

Sludge Recycle

Several brief tests were conducted to determine if recycling settled secondary sludge to the first stage of discs would enhance treatment of the wastewater. Sludge removed from the clarifier by the scoop system was collected in a drum and pumped into the first stage of discs at rates of 1% and 2% of the wastewater flow.

Table 3-3 is a summary of operating data during sludge recycle and during comparable periods without sludge recycle. The amounts of sludge recycled apparently had little effect on the concentration of mixed-liquor solids. As a consequence, there was no distinct effect on treatment efficiency. The proportion of volatile to fixed solids in the mixed liquor was unchanged by recycling sludge. The sludge recycle rates tested had no apparent effect on any system parameters.

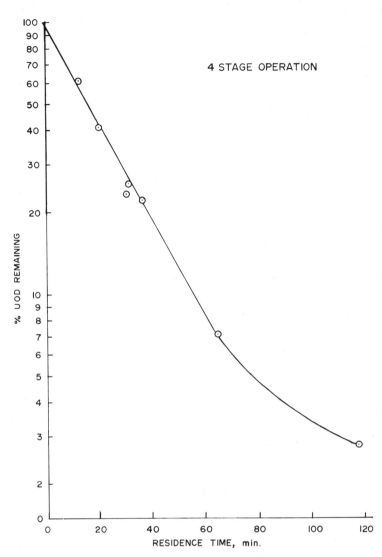

FIGURE 3-16. UOD removal kinetics as a function of retention time.

Similar work by Torpey[3] has shown that recycling sludge to achieve mixed-liquor solids concentrations as high as 2,400 mg/l has no significant effect on treatment efficiency. Since the solids attached to the rotating surface are equivalent to 10,000 to 20,000 mg/l of mixed-liquor solids, it does not appear that solids recirculation is a practical means of increasing treatment efficiency.

Torpey and his co-workers[4] have also conducted microscopic examinations of the biological growth in a multistage test unit. They revealed a succession of different types of microorganisms, starting with a predominance of zoogloeal bacteria and *Sphaerotilus* in the initial stages, which were followed by an abundant and diversified fauna consisting of free-swimming and stalked protozoa, rotifers, and nematodes in the subsequent stages. The effects of animal predation become evident in the last stages, resulting, at times, in bare spots on the surfaces. The biochemically specialized microorganisms that develop in the various stages in conjunction with changes in the substrate composition result in high efficiency of treatment.

CONTINUED PILOT PLANT STUDIES

In the summer of 1971, four rotating disc pilot plants as illustrated in Figure 3-18 were installed at

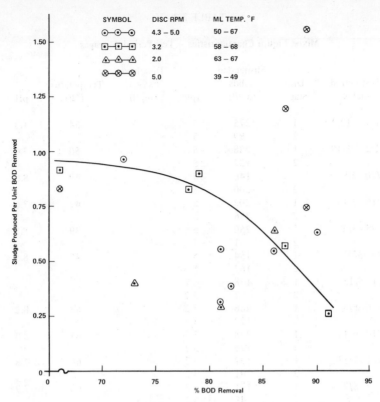

SYMBOL	DISC RPM	ML TEMP. °F
⊙—⊙—⊙	4.3 – 5.0	50 – 67
▣—▣—▣	3.2	58 – 68
△—△—△	2.0	63 – 67
⊗—⊗—⊗	5.0	39 – 49

Sludge Produced Per Unit BOD Removed

% BOD Removal

FIGURE 3-17.   Sludge production results.

the village of Pewaukee, Wisconsin, sewage treatment plant for further process development work.[5] The discs were contained in an approximately 2-ft-diameter-semicircular tank. Four cross-tank bulkheads divided the tank lengthwise into an influent chamber and four disc stages of equal size. The rotating assembly consisted of a single shaft containing a rotating bucket feed mechanism and four stages of discs. The discs were fabricated from expanded-polystyrene beads and were 23 in. in diameter and $\frac{3}{8}$ in. thick. The assembly was rotated by a 0.1-hp electric motor at 11 rpm or a peripheral velocity of 66 ft/min.

Effluent from the primary clarifier of the Pewaukee treatment plant was used as a wastewater source for the tests. Primary treated wastewater flowed by gravity from a constant-head feed box into the influent chamber of the test unit. A constant depth of wastewater was maintained in the influent chamber, so that the rotating bucket mechanism would pump wastewater into the first stage of discs at a uniform rate. Wastewater flowed from stage to stage through 1-in.-diameter submerged orifices in the bulkheads; the direction of wastewater flow was parallel to the shaft.

The units were tested without secondary clarifiers because of the difficulty in operating small-scale clarifiers. Four grab samples of the effluent were taken over a 24-hr period, settled for 30 min, and composited for analysis; 24-hr composite samples were taken of the primary clarifier effluent. The pilot plants were operated for a 1- to 2-week period under each set of operating conditions, and the test data were averaged for the following correlations. Test results from the 2-ft-diameter pilot units were also compared with data from the previously discussed tests on a 5.7-ft-diameter unit; this unit was operated in a similar fashion, except that it had a secondary clarifier and both influent and effluent samples were 24-hr composites.

Retention Time

An important factor affecting performance of the rotating contactor process is the retention time of the wastewater within the tanks containing the media. At a given hydraulic loading, measured as

TABLE 3-2

Mixed Liquor Characteristics – Test Period Averages

Test period dates	Disc stage	Suspended solids (mg/l)	rpm	Dissolved oxygen (mg/l)	Temperature (°F)	pH
11/13–12/19	1	275	5		54	7.3
	2	99	5			
12/23–1/19	1	228	5		50	7.6
	2	122	5			
1/20–2/9	1	146	5		49	8.0
	2	200	5			
2/10–2/13	1	202	5		42	7.6
	2	256	5			
2/25–3/2	1	250	5		39	8.5
	2	131	5			
3/3–3/9	1	154	5		45	8.5
	2	167	5			
4/1–5/15	1	138	3.2			
	2	97	3.2			
5/20–6/5	1	268	3.2		58	8.2
	2	123	3.2			
6/10–6/19	1	238	3.2		61	8.0
	2	126	3.2			
6/22–7/17	1	132	3.2		66	7.8
	2	91	3.2			
7/22–8/7	1	171	3.2		68	7.9
	2	49	3.2			
8/12–8/17	1	128	4.3	1.7	67	8.0
	2	72	4.3	3.6		
8/18–8/21	1	196	4.6	3.3	66	7.7
	2	89	4.6	3.7		
8/24–8/29	1	167	4.7	3.8	67	7.6
	2	96	4.7	5.1		
8/31–9/11	1	170	2.0	1.0	66	7.6
	2	50	2.0	2.6		
9/14–9/25	1	139	2.0	2.0	64	7.2
	2	84	2.0	3.8		
9/28–10/3	1	104	2.0	2.6	63	
	2	60	2.0	4.0		
10/7–10/23	1	88	4.7	3.1	63	
	2	121	4.7	4.0		

gallons per day of wastewater flow per square foot of surface area, the wastewater will have a given retention time, depending on the spacing of discs and the size of the tank containing the discs. Increasing disc spacing or increasing the tank size will increase the amount of wastewater held within the tank. This will increase the retention time at a given hydraulic loading and would be expected to increase treatment capacity. Several pilot plants like that illustrated in Figure 3-18 were tested at various disc spacings to evaluate this process variable. The results from testing at various disc spacings are correlated in Figure 3-19, using the parameter of liquid volume held within the tank per unit of disc surface area in the tank; this parameter is measured as gallons per square foot. Four volume-to-surface ratios were tested and compared with the data of Hartman,[6] i.e., the European experience at another volume-to-surface ratio. The data show that, as volume-to-surface ratio is increased from 0.067 to 0.074 to 0.085 to 0.12 gal/ft², there are increases in the treatment capacity. However, increases above 0.12 gal/ft² do not result in any further improvements. At a hydraulic loading of 1.0 gpd/ft² BOD removals of more than 96% were attained.

TABLE 3-3

Effect of Sludge Recycle on Treatment Efficiency

| Test period | Hydraulic loading (gpd/ft^2) | Mixed-liquor temp. (°F) | Sludge recycle rate (% of total flow) | Stage | Mixed-liquor suspended solids | | % Removal | | | |
					Total (mg/l)	Volatile (mg/l)	BOD	COD	SS	NH$_3$-N
2/10–2/13	1.06	42	0	1	202	160	87	69	60	27
				2	256	189				
2/16–2/20	1.06	42	1	1	230	192	84	66	81	25
				2	249	200				
2/25–3/2	0.55	39	0	1	250	220	89	70	85	94
				2	131	125				
2/23–2/24	0.47	44	1	1	88	83	85	60	85	~100
				2	140	133				
3/3–3/9	0.26	45	0	1	154	136	89	68	89	~100
				2	167	143				
3/10–3/13	0.26	40	2	1	113	90	93	88	90	~100
				2	75	61				

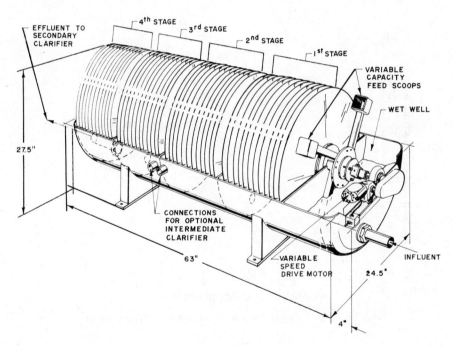

FIGURE 3-18.   Rotating disc pilot plant.

Figure 3-20 shows effluent BOD concentrations obtained for three volume-to-surface ratios tested. Data for the volume-to-surface ratio of 0.067 gal/ft^2 are not included, because the influent BOD concentrations experienced during testing of the 5.7-ft-diameter pilot plant were lower than those experienced during testing of the 2-ft-diameter pilot plants. For the volume-to-surface ratio of 0.12 gal/ft^2, an effluent BOD concentration as low as 8 mg/l was obtained at a hydraulic loading of 1.0 gpd/ft^2.

The same effect of volume to surface ratio is shown for suspended-solids removal in Figure 3-21. Increasing the volume-to-surface ratio from

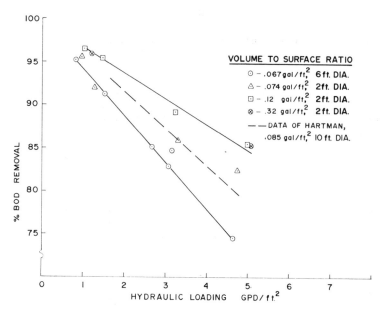

FIGURE 3-19.   Effect of disc spacing on BOD removal.

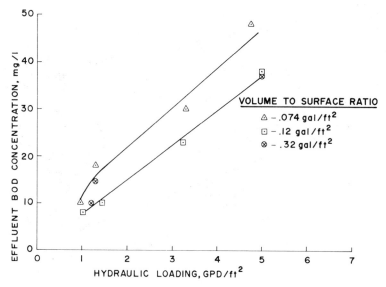

FIGURE 3-20.   Effect of disc spacing on effluent BOD concentration.

0.067 to 0.074 to 0.12 gal/ft² resulted in increased suspended-solids removal. A further increase in volume-to-surface ratio to 0.32 gal/ft², at a hydraulic loading of 5 gpd/ft², did not result in a further increase in suspended-solids removal. Figure 3-22 shows effluent suspended-solids concentrations as low as 6 to 8 mg/l at a hydraulic loading of 1.0 gpd/ft².

The rotating contactor process has been shown

to be well suited for the oxidation of ammonia nitrogen because of its staged operation. Figures 3-23 and 3-24 show the percentage of ammonia nitrogen removal and the effluent ammonia nitrogen concentration as a function of hydraulic loading for the four volume-to-surface ratios tested. These data also show 0.12 gal/ft² to be the optimum volume-to-surface ratio. Operating at volume-to-surface ratios lower than 0.12 gal/ft²

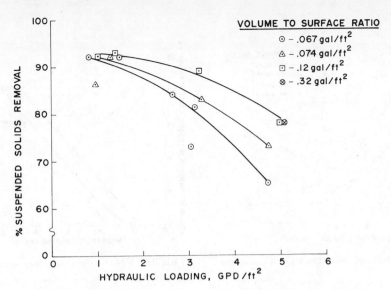

FIGURE 3-21. Effect of disc spacing on suspended-solids removal.

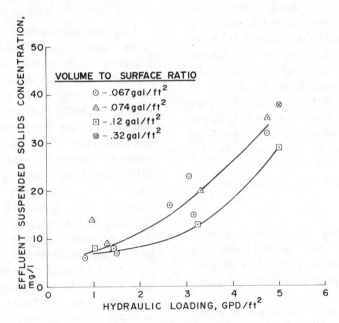

FIGURE 3-22. Effect of disc spacing on effluent suspended-solids concentration.

results in lower removals of ammonia nitrogen, and operating at ratios greater than 0.12 gal/ft^2 does not result in an increase in ammonia nitrogen removal. At a hydraulic loading of 1.0 gpd/ft^2, the process achieved virtually complete nitrification.

On the basis of these tests, it is recommended that rotating disc installations be constructed with a volume-to-surface ratio of 0.12 gal/ft^2 to make most effective use of the disc surface area. In Figure 3-23, ammonia nitrogen removals do not

loading is reduced. In Figure 3-20, 4 gpd/ft^2 corresponds to an effluent BOD concentration of approximately 30 mg/l; this, apparently, is the BOD concentration at which nitrifying bacteria can begin to compete effectively with microorganisms that chiefly oxidize carbonaceous matter.

vary significantly above a hydraulic loading of 4 gpd/ft^2; at 4 gpd/ft^2, ammonia nitrogen oxidation begins, and proceeds rapidly as the hydraulic

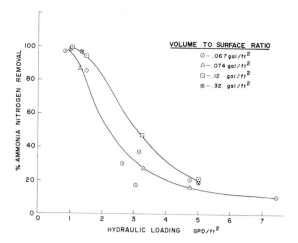

FIGURE 3-23. Effect of disc spacing on ammonia nitrogen removal.

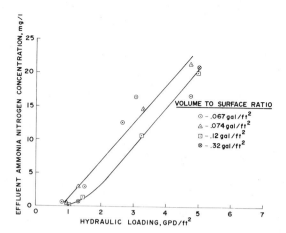

FIGURE 3-24. Effect of disc spacing on effluent ammonia nitrogen.

The predominance of a nitrifying culture in the process can be observed as a change in the color of the biomass from a dark grayish-brown to a reddish-brown or copper color. Figure 3-25* shows two of the test units that were operated indoors during the winter. The last stage (far right) of the test unit in the foreground shows the presence of some nitrifying bacteria. Figure 3-26* shows the same test units during another period of testing at a lower hydraulic loading. Here, the nitrifying bacteria advanced into the second stage, and the lighter color of the nitrifying culture is seen throughout the last three stages of treatment.

Staging

The testing described thus far was conducted on pilot plants that were divided into four individual stages of treatment. Previous testing has demonstrated four-stage treatment to be superior to two-stage treatment.

A test unit identical to that shown in Figure 3-18, but with six stages of treatment, was also tested to determine the benefits of additional staging. Figures 3-27 and 3-28 show BOD and ammonia nitrogen removal as a function of hydraulic loading for four-stage and six-stage operation. Both test units had a volume-to-surface ratio of 0.12 gal/ft². For BOD removal, six-stage operation appears beneficial for hydraulic loadings of 5 gpd/ft² and above. For lower loadings or higher degrees of treatment, four-stage and six-stage operation gave virtually identical performance. Ammonia nitrogen removal was identical for

four-stage and six-stage operation over the entire range of hydraulic loadings tested.

In a first-order system, operation in six stages rather than in four should result in increased efficiency for both BOD and ammonia removal, especially for higher degrees of treatment. The process has been shown to be first order with respect to both BOD and ammonia nitrogen removal. A reason why six-stage operation was not more effective than four-stage operation for ammonia nitrogen removal and high degrees of BOD removal can be seen in Figure 3-26. The last stage of discs of the unit in the foreground has a relatively sparse growth. This occurred because operating in four stages of treatment resulted in the development of predators, such as protozoans, rotifers, and nematodes, in the latter stages. If the number of microorganisms were the same in each stage, the benefits of staging would continue when increasing from four to six stages. However, the increased effects of predation apparently offset this benefit.

On the basis of these tests, it is generally recommended that a rotating disc installation be constructed in at least four stages of treatment to maximize the effectiveness of the surface area. If a convenient plant layout results in more than four stages in series, there may be a marginal benefit in treatment capacity.

Wastewater Temperature

The rotating contactor process is unaffected by

*See page 22A for figures 25-26.

wastewater temperatures above 55°F. For waste-water temperatures below 55°F, the treatment efficiency of the process decreases just as it does for all biological wastewater treatment processes. Figure 3-29 shows the effect of several temperature ranges and tank volume-to-surface area ratios on BOD removal. The uppermost line shows BOD removal as a function of hydraulic loading for the normal wastewater temperature range of 60 to 70°F for tank volume-to-surface area ratios of 0.12 and 0.32 gal/ft^2; the second line from the top shows the inhibition due to temperatures in the range of 39 to 50°F for the volume-to-surface

ratio of 0.32 gal/ft^2; the third line from the top shows the inhibition due to temperatures ranging from 39 to 50°F for a volume-to-surface ratio of 0.12 gal/ft^2; the bottom line shows the inhibition due to temperatures in the range of 39 to 49°F for a volume-to-surface ratio of 0.06 gal/ft^2.

Previously, it was shown that volume-to-surface ratios of 0.12 and 0.32 gave virtually identical BOD removals for the normal temperature range. However, for the low temperature range, a volume-to-surface ratio of 0.32 gal/ft^2 shows somewhat less inhibition than a ratio of 0.12 gal/ft^2. For example, a treatment plant designed for 90% overall treatment, with 86% treatment through the RBC process, would be loaded at approximately 4 gpd/ft^2 at normal wastewater temperatures. At low wastewater temperatures and a volume-to-surface ratio of 0.12 gal/ft^2, a plant would have to be loaded at approximately 2 gpd/ft^2. However, at a volume-to-surface ratio of 0.32 gal/ft^2 and low wastewater temperature, a plant could be designed for loading at approximately 3 gpd/ft^2. This shows that low wastewater temperatures can be compensated for by designing for lower hydraulic loading and/or larger tank volume-to-surface area ratio. The inhibition due to low wastewater temperature decreases as the degree of treatment increases, until, at very low loadings and very high degrees of treatment, very little inhibition occurs for either 0.32 or 0.12 gal/ft^2 volume-to-surface ratios.

Figure 3-30 shows that temperature affects

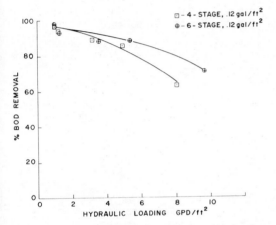

FIGURE 3-27.   Comparison of BOD removal for four-stage and six-stage operation.

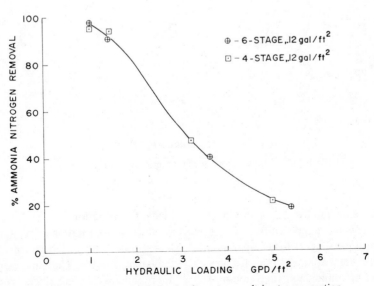

FIGURE 3-28.   Nitrification for four-stage and six-stage operation.

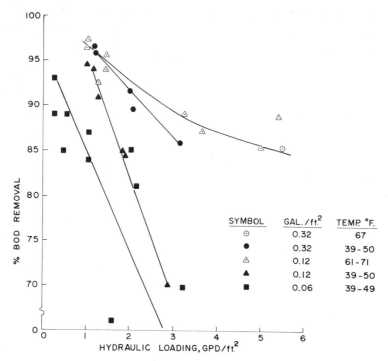

FIGURE 3-29.   Effect of wastewater temperature on BOD removal.

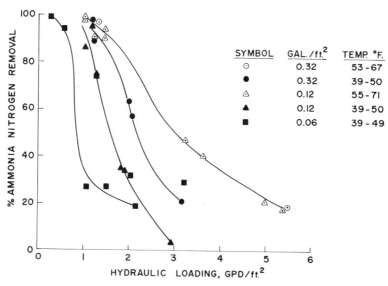

FIGURE 3-30.   Effect of wastewater temperature on nitrification.

nitrification in the same fashion it affects BOD removal. Low wastewater temperatures can be compensated for by operating at lower hydraulic loadings and/or larger tank volume-to-surface area ratios. As the degree of ammonia nitrogen removal increases, the degree of inhibition decreases.

Media Construction

Since 1960, rotating contactor plants in Europe have used discs fabricated from expanded-polystyrene beads. They are 2 or 3 m in diameter, approximately ½ in. thick, and are spaced on 1.33-in. centers. The discs are placed on metal shafts and mounted in semicircular concrete tanks. To protect the biomass from freezing temperatures and to protect the discs from damage due to wind and precipitation, the process must be covered. This can be accomplished with a simple enclosure that has windows or louvers in the sides for ventilation. There is no need to provide heat or forced ventilation for the enclosure.

In Europe, this type of construction has been competitive up to a plant size of approximately 10,000 population equivalent. Above this size, the activated sludge process has generally been more economical, in spite of the low operating and maintenance costs of the rotating contactor process. To make the latter process competitive over a wider range of plant sizes, it was necessary to develop a new disc design with a significantly lower cost and yet maintain the low power consumption.

Figure 3-31 shows a cross section of the type of media that has been developed. It is fabricated from high-density polyethylene in alternating flat and corrugated sheets, which have been thermally bonded together to form a continuous honeycomb-like structure. A stack of media with approximately 1¼ in. layer-to-layer spacing has surface area density of approximately 37 ft^2/ft^3, which is more than twice that of the European disc design. The key feature to the operation of this design is the provision of radial passages extending from the shaft to the outer perimeter of the media. This assures that wastewater, air, and stripped biomass can pass freely into and out of the media assembly. In full-scale size, radial passages are provided every 30°.

The media is available in diameters up to 12 ft and is mounted on horizontal shafts up to 25 ft long. The shaft assembly shown in Figure 3-32 has approximately 100,000 ft^2 of surface area available for developing a fixed biological growth. Figure 3-33 is a schematic drawing of the corrugated media in operation. As the media rotates into the wastewater, the wastewater flows into the radial passages and is distributed to all of the corrugations. The rotating surfaces carry the entrained wastewater to the opposite side of the tank, and as the media rotates upward, the wastewater drains out of the corrugations into the radial passages and then out into the bulk of the wastewater in the tank. At the same time, new air flows into the radial passages. Each rotation results in a complete change of both wastewater and air and creates a mixing pattern within the tank

sufficient to assure that all wastewater frequently comes in contact with the attached biomass.

Use of the higher surface density and the larger diameter have reduced the cost of rotating equipment and have also significantly reduced concrete-tankage, enclosure, and land requirements. The old polystyrene discs are fragile and susceptible to damage during assembly, shipment, installation, and erection of the balance of the treatment plant. The new construction is quite strong and resistant to damage. Shaft assemblies are shipped to treatment plant sites as shown in Figure 3-32, without any protection for the media.

Pilot Plant Tests

To determine the effectiveness of the surface area in the new design, a 2-ft-diameter prototype

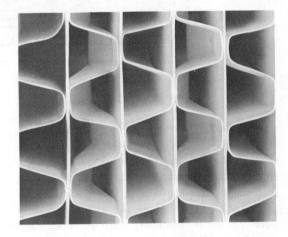

FIGURE 3-31.   Cross section of corrugated media.

FIGURE 3-32.   Corrugated media assembly.

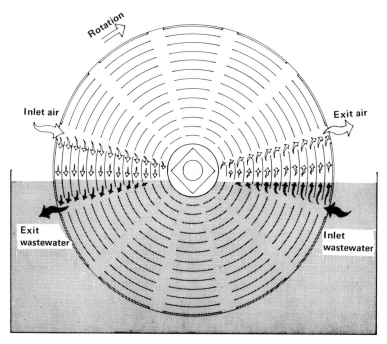

FIGURE 3-33. Schematic diagram of wastewater distribution and flow in corrugated media.

was installed at the village of Pewaukee, Wisconsin, and tested with a 2-ft-diameter pilot plant containing the expanded-polystyrene bead disc construction.[7] Construction, operation, and testing of the unit were identical to those described for the unit in Figure 3-18. The corrugated discs were 23 in. in diameter, 0.03 in. thick, and spaced on 1¼ in. centers. The assembly was rotated by a 0.1-hp electric motor at 11 rpm or a peripheral velocity of 66 ft/min.

BOD and Suspended Solids Removal

Figures 3-34 to 3-37 show the percentage of reduction and the effluent concentrations for BOD and suspended solids as a function of hydraulic loading on the disc surface area. The data show that, over the full range of hydraulic loadings tested, the surface area of the discs with formed-sheet construction performed on an equivalent basis with the surface area of the expanded-bead construction. BOD and suspended-solids removals as high as 96% and 94%, respectively, were achieved at a hydraulic loading of approximately 1.0 gpd/ft². This yielded effluent BOD and suspended-solids concentrations of 6 to 8 mg/l. At no time during the testing was there any evidence of clogging of the media. Shearing forces exerted

on the attached culture, as it was passed through the wastewater at a peripheral velocity of 66 ft/min, kept the concentric corrugations and radial passages free of excess biomass.

Figures 3-38 and 3-39 show the percentage of ammonia nitrogen removal and the effluent ammonia nitrogen concentration as a function of hydraulic loading. Again, the surface area of the two types of disc designs are interchangeable over the entire range of loadings investigated. Ammonia nitrogen removal of more than 95% and effluent concentration of less than 1.0 mg/l ammonia nitrogen were attained at a hydraulic loading of 1.0 gpd/ft².

Nitrogen Removal Characteristics

The development and predominance of ammonia-oxidizing organisms in the process has been found to be primarily a function of BOD concentration. At high BOD concentrations, these organisms cannot compete with the more rapidly growing carbon-oxidizing organisms and are diluted out of the process through population dynamics. As carbonaceous matter is removed by biological treatment, and as the BOD concentration approaches 30 mg/l, nitrifying organisms begin to establish themselves in the process.

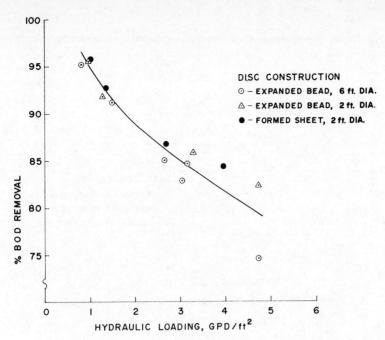

FIGURE 3-34.   Comparison of BOD removal for different media.

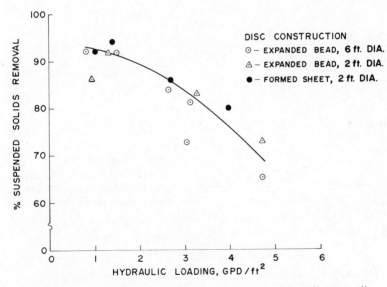

FIGURE 3-35.   Comparison of suspended-solids removal for different media.

Ammonia oxidation begins and increases rapidly as the BOD concentration is further reduced. When the BOD concentration reaches 8 to 10 mg/l, ammonia oxidation is virtually complete (less than 1.0 mg/l). This relationship is shown in Figure 3-40.

Test data from several units operated under various conditions is included are Figure 3-40 to more firmly establish the relationship. These data also show the relationship to be independent of influent BOD concentration, fresh or septic wastes, wastewater temperature, rotational disc velocity, and number of stages of discs (when four or more).

The rotating contactor process removes Kjeldahl nitrogen in proportion to ammonia nitro-

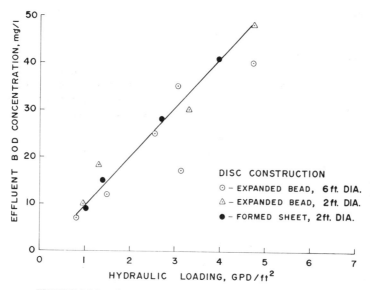

FIGURE 3-36.   Comparison of effluent BOD for different media.

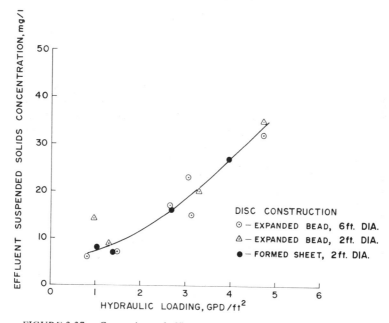

FIGURE 3-37.   Comparison of effluent suspended-solids for different media.

gen removal. Figure 3-41 shows that effluents of 1 to 2 mg/l are obtained at essentially complete ammonia nitrogen oxidation. This relationship permits determination of the amount of organic nitrogen in an effluent for various degrees of ammonia nitrogen removal. The mechanism of organic nitrogen removal is primarily through cell synthesis, however, flocculation and settling, and some hydrolysis to ammonia nitrogen are probably occurring.

Varying Hydraulic Loads

All of the development work described thus far has been at relatively constant wastewater flows. To determine the stability of the system under varying flow conditions the following tests were conducted.[8]

Wastewater flow from many municipalities often follows a pattern similar to that shown in Figure 3-42 where there are periods of increasing, decreasing, and relatively constant flow. This flow

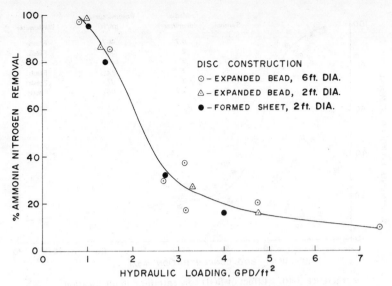

FIGURE 3-38.   Comparison of nitrification for different media.

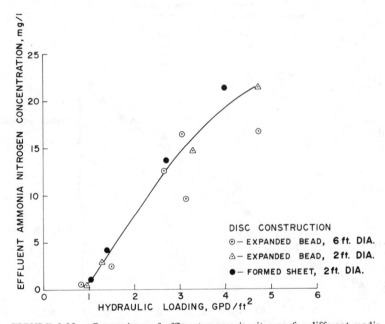

FIGURE 3-39.   Comparison of effluent ammonia nitrogen for different media.

pattern and a constant COD concentration were used to simulate rotating contactor operation under conditions of cyclic flow.

The test systems used consisted of 3-ft diameter aluminum discs arranged in two separate stages each containing 87 discs. A synthetic waste was used consisting of dairy solids and inorganic nutrients. Performance under these flow conditions and expected performance under uniform

flow conditions are presented in Figure 3-43. Stage 1 showed a variation in percent COD reduction corresponding to uniform flow performance which indicates that its overall performance is not reduced by varying the loading. In fact, the overall performance, i.e., the average percent reduction for the entire flow cycle, is somewhat better than would be expected for steady state flow.

Performance of Stages 1 and 2 in series did not

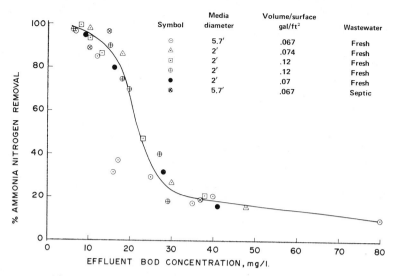

FIGURE 3-40.   Effect of BOD concentration on nitrification.

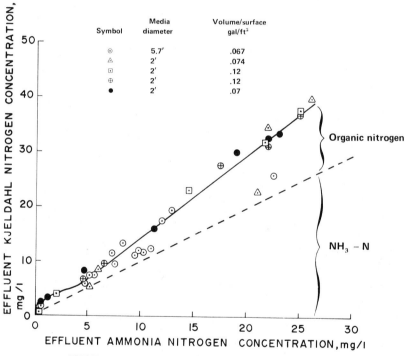

FIGURE 3-41.   Distribution of effluent nitrogen content.

follow the pattern for uniform flow operation very closely. It varied generally between 70 and 80% COD reduction which is somewhat higher than the overall performance for uniform flow operation. The units in series did not obtain the high percent reduction experienced at uniform flow for the low flows period; this may be due to the fact that during periods of high loading, bacteria can make

use of a limited capacity to store substrate to be metabolized when the loading is reduced. This results in better performance during a period of high loading and poorer performance during a subsequent period of lower loading. The result is a fairly constant performance over the loading cycle. If the same number of discs had been divided into more than two stages, the performance may have

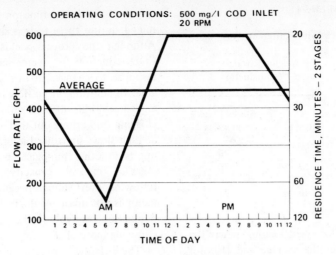

FIGURE 3-42. Varying flow pattern.

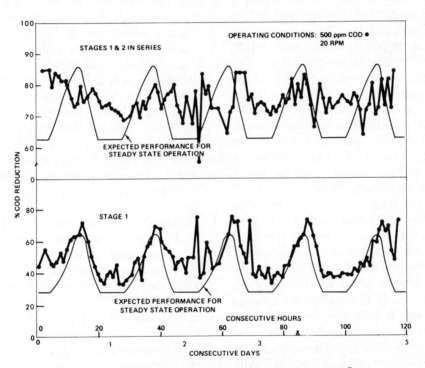

FIGURE 3-43. RBC system performance with varying wastewater flow.

been more constant over the flow pattern. Previous steady state testing under similar conditions yielded 39% COD reduction in Stage 1 and 70% COD reduction over both stages. Calculating a weighted average, percent COD reduction (weighted on the basis of flow) for the entire 5 days of the simulation yielded 46% COD reduction in Stage 1 and 73.2% COD reduction over both

stages (see Table 3-4). This slightly better reduction indicates that at least equal, if not better performance is obtained when wastewater flow varies in a cyclic pattern.

A German investigator, Professor Franz Pöpel of the Technical University of Stuttgart, West Germany, has conducted similar studies[9] on the process and his results are presented in Figure

TABLE 3-4

RBC System Performance with Different Wastewater
Flow Patterns

Flow patterns	Weighted average, % COD reduction	
	Stage 1	Stages 1 and 2 in series
Steady state	37	70
For average flow (450 gph, $\overline{T}$ = 27 min)	39	70
For cyclic flow	46	73

3-44. They show that for cyclic loading, where the peak load is four times the average and lasts for 2.5 hr the BOD removal efficiency is greater than it would be for a uniform load. As the duration of the peak is increased to 5 hr the improvement over uniform loading is decreased. Further increases in the duration of the peak load could result in BOD removals which are less that of uniform loading.

One problem which seriously affects municipal wastewater treatment plants particularly during wet weather conditions is hydraulic surges. To determine the stability of the rotating contactor process under this flow condition a different test system was used. Figure 3-45 is an overall view of the test setup. This rotating contactor unit is approximately 12 ft long, 1½ ft wide, and 1 ft high with an operating capacity of approximately 60 gal. The test unit contains 320, 1-ft-diameter aluminum discs arranged in a series of 10 parallel rows. Figure 3-46 is a flow sheet for the system. The synthetic wastewater was the same as that for the previous tests. To ensure an adequate supply of tap water at near room temperature, a 1,300-gal reservoir was used to store water for the hydraulic surge tests. Water in this reservoir was pumped to the mixing drum to dilute the concentrated sewage to the desired concentration of 500 mg/l COD. Simulating hydraulic surges to the system was done to determine its COD removal capacity during extremely high loading and to determine how quickly it can return to steady state efficiency once the surge has ceased. Figure 3-47 shows the response to the changes in flow rate. It took the system an hour to recover steady state efficiency once the high flows ceased and steady state flow was reinstated.

Figure 3-48 shows the increased COD removal capacity obtained during the high flows. The effect of residence time on COD removal determined during these tests is shown in Figure 3-49. Although the process yields decreasing percent COD reduction for decreasing residence time, its response to changes in flow is rapid and predictable. In addition, at no flow rate tested were there any deleterious effects on the biomass such as clogging or washing out that are often experienced at high flows by other biological processes. These upsets in other biological processes reduce their organic removal capacity both during and following overloading. Because the rotating contactor is not upset by the high flows, its microbial population responds to the higher organic load by removing more of it.

The rotating contactor process is not upset by hydraulic surges, because it has a captive microorganism population. For residence times as low as 3 min. no measurable stripping or sloughing of biomass solids occurs. The increased organic loading accompanying a hydraulic surge is, to a significant degree, absorbed by the large microbial population, and return to steady state effectiveness following the surge is rapid and complete.

Another German investigator, E. Märki, has conducted similar investigations[10] on German municipal wastewater. His results are presented in Table 3-5. For two different steady state conditions he subjected a rotating contactor system to a tenfold increase in hydraulic load for 8 hr each week and measured its BOD removal before, during, and following the overload. Significant BOD removals were obtained during the overload period and BOD removal the following day was rapidly approaching the previous steady state level. These data and the results reported in a previous paper by Welch[11] are compared to that developed during these tests in Figure 3-50. Considering the increased effectiveness of multistage systems for a given residence time, these data of all three investigations are in agreement.

Enriched Oxygen Atmosphere

Torpey[3] has done extensive work on the use of enriched oxygen atmosphere with rotating contactors. His findings indicate that at high loadings, when there may be an oxygen limitation on the initial stage of media, the use of enriched oxygen can approximately double BOD removal rates. At the present time, the relative costs of rotating contactor equipment and oxygen generation and utilization favor the use of additional rotating

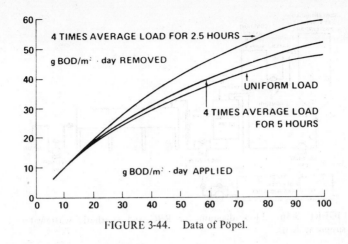

FIGURE 3-44.   Data of Pöpel.

FIGURE 3-45.   Experimental system for hydraulic surge tests.

contactor equipment. However, there may be applications such as treatment of concentrated wastes and plant sites with limited land area where use of enriched oxygen will be cost effective.

RECENT DEVELOPMENTS

Three of the principal reasons for the growing acceptance of the rotating contactor are its simplicity, low energy consumption. and ease of achieving nitrification. Two new developments will enhance these benefits.

During the past two years a new version of the

corrugated polyethylene media has been tested which has approximately a 50% increase in surface area density. This new media has proven to be applicable to nitrification, where the thinner biomass allows the spacing between adjacent layers of media to be significantly reduced. The new media construction has approximately 150,000 ft² of available surface area per shaft. This results in a reduction of one third in the number of shaft assemblies required for nitrification applications. Very few other changes in facilities and ancillary equipment are required. Therefore, because of its modular construction, the total installed cost,

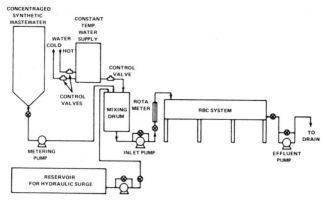

FIGURE 3-46. Flow diagram for RBC and synthetic wastewater supply systems.

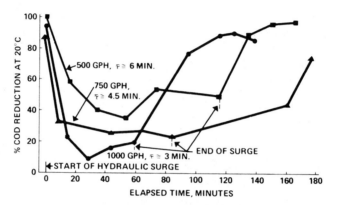

FIGURE 3-47. Effect of hydraulic surge on COD removal efficiency.

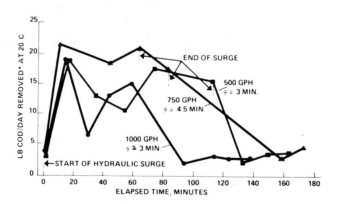

*TO CONVERT TO LB COD/DAY/1000 FT³ OF DISC VOLUME MULTIPLY BY 95

FIGURE 3-48. Effect of hydraulic surge on COD removal capacity.

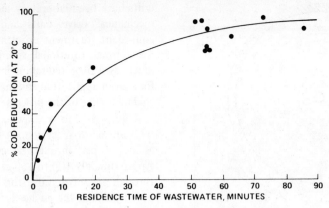

FIGURE 3-49. Effect of residence time on BIO-DISC system performance.

TABLE 3-5

Data of E. Märki

	Steady state	For 8 hr each week	Following day (24 hr later)
Residence time, min	120	12	120
BOD reduction, %	92.5	46.9	85.2
Residence time, min	300	30	300
BOD reduction, %	94.6	65.0	77.6

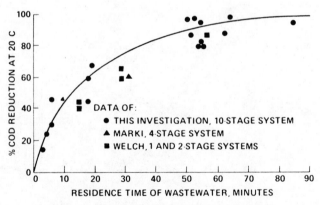

FIGURE 3-50. Comparison with Figure 3-49.

including enclosures, tankage, piping, electrical, etc., is reduced between 25 and 30% per unit of surface area.

The second new development is an alternative drive system for the shaft assemblies. This is shown schematically in Figure 3-51. It consists of plastic cups which are welded around the outer perimeter of the media and over the entire length of the contactor. The media assembly is installed in the tank in the normal manner. A small air header is placed below the media and releases air at a low pressure into the attached cups. The captured air exerts buoyant force, which in turn exerts a torque on the shaft sufficient for rotation. Because air cups are not placed over those portions of the media containing a radial passage opening,

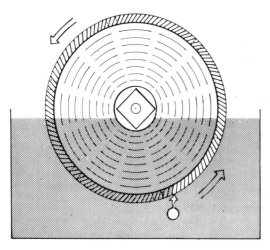

FIGURE 3-51.   Air drive system.

units have been laboratory tested side by side with mechanical drive units and have demonstrated equivalent treatment efficiency at as little as half the normal rotational speed. This results in as much as a 50% reduction in power consumption for a given application and makes the energy saving characteristics of the process even more dramatic.

Figure 3-52 is a drawing of a full-scale plant with an air drive system. A blower delivers air at about 3 psi through a common air header. Approximately 100 to 150 cfm of air is withdrawn through a control valve for each shaft and distributed through the air header beneath the rotating media. The simplicity of this drive system is very important for larger treatment plans employing a large number of shafts. Rather than having many individual mechanical drive systems, there are now only one or two central blowers delivering air through common headers to drive a large number of shafts. This significantly reduces overall maintenance requirements and allows individual adjustment of rotational speed for each shaft. The treatment plant operator then can "fine tune" his installation to meet his effluent standards while minimizing energy consumption.

approximately 20% of the released air escapes into the radial passages and flows upward into the corrugated sections. The supplementary aeration achieved from this is sufficient to allow a reduction in rotational velocity of the media while still achieving the same degree of treatment. Full-scale

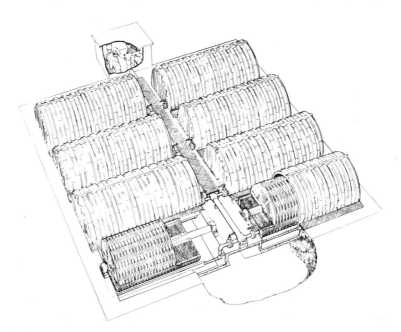

FIGURE 3-52.   Rotating biological contactor treatment plant with air drive.

REFERENCES

1. Application of Rotating Disc Process to Municipal Wastewater Treatment, U.S. Environmental Protection Agency Water Pollution Control Research Ser., Project 17050 DAM, November, 1971.
2. **Antonie, R. L.,** Application of the BIO-DISC Process to treatment of Domestic Wastewater, paper presented in part at the 43rd Ann. Water Pollut. Control Fed. Conf., Oct 4–9, 1970, Boston, Mass.
3. **Torpey, W. N.,** Rotating Biological Disk Wastewater Treatment Process – Pilot Plant Evaluation, Rutgers University, Research and Development Office, EPA Project 17010 EBM, August 1972.
4. **Torpey, W. N. et al.,** Rotating disks with biological growths prepare wastewater for disposal or reuse, *J. Water Pollut. Control Fed.,* 43, 2181, 1971.
5. **Antonie, R. L.,** Factors Affecting BOD Removal and Nitrification in the BIO-DISC Process, paper presented at the Central States Water Pollut. Control Assoc. Annu. June 14–16, 1972, Milwaukee, Wis.
6. **Hartmann, H.,** Der Tauchtroptkörper, *Oesterr. Wasserwirtsch.,* 17, 264, 1965.
7. **Antonie, R. L.,** Three-step Biological Treatment with the BIO-DISC Process, paper presented at the N.Y. Water Pollu. Control Assoc. Spring Meeting, June 12–14, 1972, Montauk, N.Y.
8. **Antonie, R. L.,** Response of the BIO-DISC Process to Fluctuating Wastewater Flows, Proc. 25th Purdue Industrial Waste Conf., May 5–7, 1970, 427, W. Lafayette, Ind.
9. **Pöpel, F.,** Construction, degradation capacity and dimensioning of rotating biological filters, *Eidg. Tech. Hochsch. Zurich-Fortbildungsk. EWAG,* 26, 394, 1964.
10. **Märki, E.,** Results of experiments by EWAG with the rotating biological filter, *Eidg. Tech. Hochsch. Zurich-Fortbildungsk. EWAG,* 26, 408, 1964.
11. **Welch, F. M.,** Preliminary Results of a New Approach in the Aerobic Biological Treatment of Highly Concentrated Wastes, Proc. 23rd Purdue Ind. Waste Conf., May 6–8, 1968, W. Lafayette, Ind., 428; *Water Wastes Eng.,* 6(7), D–12, 1969.

Chapter 4

DESIGN CRITERIA – DOMESTIC WASTEWATER TREATMENT

DESIGN CRITERIA

Process design criteria have been established through extensive pilot plant testing in the U.S. since 1965, and from operating experience since 1969.

Hydraulic Loading

The rotating contactor process has been found to demonstrate first order kinetics for the removal of carbonaceous BOD, oxidation of ammonia nitrogen, and removal of ultimate oxygen demand. This means that at a specific hydraulic loading, a specific percentage removal of BOD will occur independent of influent BOD concentration. Because of this, the primary design criterion is hydraulic loading, and not organic loading as is often practiced with the activated sludge and trickling filter process. To simplify design calculations, hydraulic loading is expressed as flow per unit time per unit of surface area covered by biological growth, or gallons per day per square foot (gpd/ft^2). It might seem that retention time would be the best means of determining hydraulic loading. However, since actual retention time can be calculated only by estimating the void volume of biomass covered media, and cannot be directly translated into a requirement for a specific amount of rotating equipment, hydraulic loading on the biomass covered surface is used for determining equipment requirements. Therefore, the main effort associated with design and selection of equipment for any wastewater treatment application is to determine the requirement for growth covered surface area.

Kinetic evaluation of the rotating contactor process, using unit mass removal rates $\left(\frac{\text{mg/l BOD}}{\text{mg/l VSS}}\right)$ such as is done for the activated sludge process, is not possible because of difficulty in determining the amount of biomass in the system. As a practical matter, however, this type of evaluation is not really valid for the rotating contactor process. No practical means exist to control the amount of biomass. The amount and type of biomass that develops in the successive stages of treatment will be that best suited to treating the wastewater, therefore, control is neither necessary nor desirable.

Sludge age, determined by comparing an estimated amount of attached biomass to the net sludge generation in the effluent, indicates a range from 3 days when a concentrated industrial waste is being treated to about 1 month or more when a secondary effluent is being nitrified. Sludge age is not a factor, because it is exceptionally long.

The food:microorganism (F:M) ratio has been shown to be approximately equivalent to an activated sludge system[1] and is important in process operation. Because the F:M ratio cannot be controlled, however, it does not provide a practical basis for design. The process has exhibited first-order characteristics in oxidizing BOD and ammonia nitrogen, so hydraulic loading has been adopted as the principal design criterion. Although this represents an empirical approach to design, it is easy to understand and to use and it has achieved reproducible results.

Staging and Plant Arrangement

The arrangement of media in a series of stages has been shown to significantly increase treatment efficiency. This occurs for two reasons. First is the development of specific microbial cultures in the successive stages of media which are adapted to the wastewater characteristics in each stage. With domestic wastewater, the latter stages of media develop nitrifying organisms which oxidize ammonia nitrogen (see Chapter 3 for detailed discussion). Secondly, because the process exhibits first order kinetics, the improved residence time distribution (i.e., more closely approaching "plug flow"), obtained with staging, increases the BOD removal rate. In "plug flow" operation, organisms in the first stage of media are exposed to a high BOD concentration and respond by removing BOD at a high rate. As the BOD concentration decreases from stage to stage, the rate at which the organisms remove BOD also decreases. The average BOD removal rate is greater than if all the media were in a single completely mixed stage where all organisms are exposed to a relatively low BOD concentration. Thus, it is desirable to construct rotating contactor plants in at least four stages to most effectively utilize the surface area. For treatment plants requiring many shafts of media, convenient plant layout often calls for more than four stages in series. This can be done for most domestic wastewater applications without fear of overloading the first stage and will result in a slight increase in treatment efficiency over four-stage operation (see Chapter 3 for details). Treatment plants requiring four or more media assemblies are arranged so that each shaft is an individual stage of treatment. The shafts are arranged in series and the wastewater flow is perpendicular to the shafts. For plants where fewer than four shafts are required, they can be arranged parallel to one another. Each tank containing a shaft is divided into stages with cross-tank bulkheads along its length, and wastewater flow is parallel to the shaft. Each bulkhead has a submerged orifice, and each section of media between bulkheads acts as a separate stage of treatment. Tests have shown that each stage is completely mixed, and that there is no difference in treatment capacity when wastewater flow is either parallel or perpendicular to the shaft (see Chapter 7 for details on equipment arrangements).

Effluent BOD Characteristics

Effluent from a rotating contactor unit providing nitrification contains nitrifying organisms. Because of this, significant nitrification occurs during a 5-day BOD test on the effluent. In BOD tests where allylthiourea was added to dilution water to suppress nitrification, it was shown that an effluent of 30 mg/l total BOD_5 or less could contain as much as 50% nitrogenous BOD. This relationship is valid for effluents as low as 10 mg/l total BOD_5. Below this BOD level, nitrification is essentially complete and the proportion of carbonaceous BOD increases. Total BOD_5 removals of 85% and 90% then correspond to carbonaceous BOD_5 removals of as much as 90% to 95%, respectively. This characteristic is an important consideration when evaluating operating data from any treatment plant employing a biological treatment process. If the process is nitrifying or on the verge of nitrifying, additional 5-day BOD tests should be performed with a nitrification suppressant to determine its true treatment efficiency.

Rotating contactor process effluents of 25 mg/l BOD and less have approximately one half of the remaining BOD as suspended solids. Therefore, tertiary filtration can achieve an additional 50% BOD removal to produce final effluents as low as 5 mg/l BOD and less.

Media Rotation

Rotational velocity of the media is also an important design criterion. Testing of various diameter media indicates that a fixed peripheral velocity can be used to determine the required rotational velocity for any diameter.

Rotational velocity affects wastewater treatment in several ways: it provides contact between the biomass and the wastewater, it aerates the wastewater, and it provides energy to thoroughly mix the wastewater in each stage. Increases in rotational velocity increase the effect of each of these factors. However, there is an optimum rotational velocity above which further increases in these factors no longer increase treatment levels. This optimum velocity will vary with wastewater BOD concentration, i.e., the optimum velocity is higher for concentrated industrial wastes and lower for domestic waste. Also, the optimum rotational velocity will decrease from stage to stage as the BOD concentration decreases from stage. It has been found that when all stages of discs in the plant rotate at the same velocity, the optimum peripheral velocity for domestic wastewater is 60 ft/min. This is true for BOD removal and nitrification. See Chapter 3 for more details.

Since power requirements increase exponentially with increases in media velocity, there is a practical upper limit of rotational velocity used for industrial waste treatment. The ability to maintain a large attached culture is not a factor in selecting rotational velocity. Pilot plant testing at velocities well above practical limits on the basis of power consumption (300 to 400 ft/min)[2,3] have shown no loss in the amount of biomass.

The direction of media rotation has no effect on treatment efficiency and is not a factor in selecting rotational velocity. In a multishaft installation with flat bottom tankage, the immersed portion of the media is rotated in the opposite direction as the wastewater flow to minimize any possible short-circuiting along the bottom of the tank.

Tank Volume

An important factor affecting performance of the process is the retention time of the wastewater within the tanks containing the media. At a given hydraulic loading, as gpd/ft², the wastewater will have a given retention time depending upon the void fraction of the media, and the size of the tank containing the media. Increasing void fraction of the media or increasing the tank size will increase the amount of wastewater held within the tank. This will increase the retention time at a given hydraulic loading and will, therefore, improve performance. Extensive testing using various void fractions and tank sizes has led to the conclusion that there is an optimum tank volume which maximizes the treatment capacity of the growth covered surface. For purposes of plant design, this tank volume is measured as wastewater volume held within a tank containing a shaft of media per unit of growth covered surface on the shaft, or gallons per square foot (gal/ft^2). The optimum tank volume determined when treating domestic wastewater up to 300 mg/l BOD is 0.12 gal/ft^2, which takes into account wastewater displaced by the media and attached biomass. The use of tank volumes in excess of 0.12 gal/ft^2 does not yield corresponding increases in treatment capacity when treating wastewaters in this concentration range. All design relationships and process layouts described here and in subsequent chapters use this tank volume.

Wastewater Temperature

Wastewater temperature affects rotating contactor performance just as it does all biological wastewater treatment processes. Wastewater temperatures between 55 and 85° F have no effect on process performance. When wastewater temperatures decrease below 55° F, the treatment efficiency will also decrease.

If wastewater flows are sufficiently lower during periods of low wastewater temperatures, then treatment efficiency will be maintained. In cases where low wastewater temperature is due to sewer infiltration or run-off from rainfall, the conditions of lower temperature will not coincide with lower flows, consequently, treatment efficiency will not be maintained. Infiltration, however, will generally dilute the raw wastewater so that while percentage removal may decrease, the effluent concentration may not be materially affected. Also, discharge standards for a receiving body may not be as stringent under cold weather conditions. If it is required that a given percentage treatment or maximum effluent quality be maintained under all conditions, then it will be necessary to design the plant to offset the effect of the low wastewater temperature.

Enclosures

Year-round operation in northern climates requires that rotating contactor plants be covered to protect the biological growth from freezing temperatures and avoid excessive loss of heat from the wastewater. Some industrial wastes have inherent odor problems. The enclosure for the process will facilitate odor control measures. Installations in southern climates, or installations in northern climates which operate during the warmer seasons only (such as recreational areas), need not be totally enclosed except for aesthetic reasons. Wind will not damage the media and precipitation can remove only a small amount of the biomass from the corrugated media.

Enclosures can be constructed of any suitable corrosion resistant material. Heating or forced ventilation are not necessary. Windows or simple louvered mechanisms which are opened in the summer and closed during the winter provide adequate ventilation. Air within the enclosure is at a temperature approximately equal to that of the wastewater. At very low ambient air temperatures, the high humidity within the enclosure will result in condensation on the walls and ceiling. To minimize corrosion within the enclosure and increase operator comfort, the condensation can be eliminated by insulating the enclosure or heating the air within the enclosure. Because condensation will occur only during cold weather, heating will sometimes be more economical.

To reduce the cost of enclosing a rotating contactor plant, molded plastic covers with or without thermal insulation are also available. This type of enclosure minimizes the area to be covered and also eliminates the need for the operator to enter the enclosure. This eliminates the need for heating or lighting the enclosure or providing for other operator comforts. Sufficient ventilation is provided by openings in the junctions of cover sections.

DESIGN CALCULATIONS

Surface Area Determination

The following design procedure applies to the majority of domestic wastewater treatment applications. For a given degree of treatment, the information in Figure 4-1 and Table 4-1 is used to determine the required hydraulic loading on the process. When the hydraulic loading rate has been determined, the surface area can be computed as shown in Example #1 below.

Example #1 — It is necessary to remove 90% of the BOD from a 1.0-mgd wastewater flow containing 200 mg/l BOD. The effluent from primary

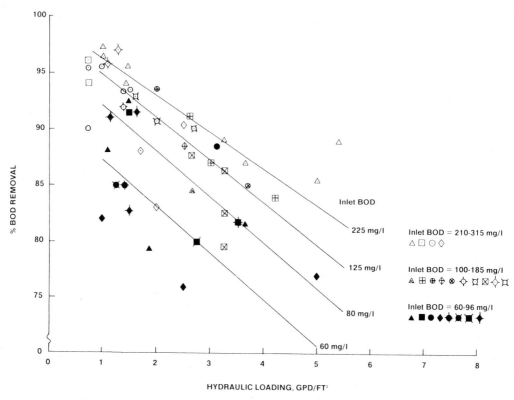

FIGURE 4-1.   Design correlation for BOD removal in municipal wastewater treatment.

TABLE 4-1

RBC Process Municipal Wastewater Treatment BOD Removal

0.5-m-diameter pilot plants

Inlet BOD = 210−315 mg/l
  △ −Pewaukee, Wis.   210−260 mg/l
  ⬠ −Spencer, Ia.   210−290 mg/l
  ◇ −Bemidji, Minn.   220−250 mg/l
  ○ −Luverne, Minn.   315 mg/l

Inlet BOD - 100−175 mg/l
  ◇ −Bemidji, Minn.   157 mg/l
  ○ −Spring City, Pa.   126 mg/l
  ○ −Luverne, Minn.   175 mg/l
  ○ −Washington, Pa.   100−125 mg/l
  ⬠ −Philadelphia, Pa.   110−150 mg/l
  △ −Pewaukee, Wis.   110 mg/l
  ○ −Tallahassee, Fla.   110−150 mg/l

Inlet BOD = 34−92 mg/l
  ▲ −Pewaukee, Wis.   34−92 mg/l
  ● −Tallahassee, Fla.   83 mg/l
  ■ −Luverne, Minn.   90 mg/l
  ◆ −Shelbyville, Ind.   57−64 mg/l
  ● −Washington, Pa.   80 mg/l

Full-scale plants and 3.2-m-diameter pilot plants

Inlet BOD = 100−185 mg/l
  ⬠ −Philadelphia, Pa.   106−124 mg/l
  ◇ −Gladstone, Mich.   103−122 mg/l
  ⬠ −Woodland, Wash.   181 mg/l

Inlet BOD = 60−96 mg/l
  ● −Omaha, Neb.   60 mg/l
  ■ −Niagara, Wis.   83 mg/l
  ◆ −Gladstone, Mich.   78−96 mg/l

treatment is expected to have a 150 mg/l BOD concentration. To obtain a 20 mg/l BOD secondary clarifier effluent, the rotating contactor process must then achieve 86.7% BOD removal. From Figure 4-1, a 150 mg/l primary effluent BOD and 86.7% BOD removal requires a hydraulic loading of 3.5 gpd/ft². The process surface area then required is:

$$\frac{\text{Flow}}{\text{Hydraulic Loading}} = \frac{1,000,000 \text{ gpd}}{3.5 \text{ gpd/ft}^2} = 286,000 \text{ ft}^2$$

Effect of BOD Concentration

The rotating contactor process is approximately first order with respect to BOD removal; i.e., for a given hydraulic loading (or retention time) a specific percent BOD reduction will occur, regardless of the influent BOD concentration. However, BOD concentration does have a moderate effect on degree of treatment as indicated by the family of curves in Figure 4-1. This is due to the limit on effluent BOD concentration achievable by biological treatment and the ability to settle all of the

biological solids produced from treatment. This effect of a limiting effluent concentration increases as the degree of treatment increases. Another factor contributing to lower percentage BOD removals at low influent BOD levels is the onset of nitrification in upstream stages. Once nitrifying organisms predominate, the rate of carbon oxidation slows down significantly.

Temperature Correction

When the temperature of the wastewater is 55°F or above, no further design calculations are necessary. However, when climatic conditions significantly reduce wastewater temperature for prolonged periods, it may be necessary to adjust the design to a lower hydraulic loading to achieve the desired degree of treatment during these low temperature periods.

The effect of wastewater temperature on BOD removal can be seen from the data in Figure 4-2 and Table 4-2. As the temperature decreases below 55°F, BOD removal decreases. The effect of

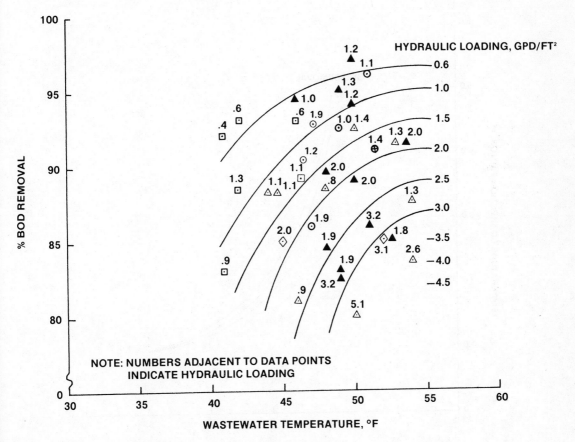

FIGURE 4-2.   Effluent of wastewater temperature on BOD removal.

51

temperature is somewhat greater for lower degrees of treatment and decreases slightly as the degree of treatment increases. At high degrees of treatment where there is a relatively long wastewater retention time, temperature has slightly less effect on treatment efficiency. This can be seen in Figure 4-3, which was derived from Figure 4-2 by calculating the ratio of required hydraulic loadings for various temperatures and degrees of treatment.

TABLE 4-2

Municipal Wastewater Treatment at Low Wastewater Temperatures

0.5-m-diameter pilot plants	Full-scale plants and 2.0-m-diameter pilot plants
Inlet BOD = 80–180 mg/l Δ – Pewaukee, Wis. ⊕ – Washington, Pa.	Inlet BOD = 80–180 mg/l ○ – Gladstone, Mich. □ – Brewster, N.Y. ◇ – Pewaukee, Wis.
Inlet BOD = 210–315 mg/l ▲ – Pewaukee, Wis.	

Example #2 – For the same conditions described in Example #1, the wastewater is expected to be 45°F, and it is still necessary to maintain the same degree of treatment. From Figure 4-3, the temperature correction factor, for 45°F and 86.7% BOD removal, is interpolated to be 2.1. The adjusted hydraulic loading to obtain 86.7% BOD removal with 45°F wastewater is:

$$\frac{\text{Hydraulic Loading } (T \geq °F)}{\alpha} = \frac{3.5 \text{ gpd/ft}^2}{2.1} = 1.67 \text{ gpd/ft}^2$$

The surface area requirement is then

$$\frac{1,000,000 \text{ gpd}}{1.67 \text{ gpd/ft}^2} = 600,000 \text{ ft}^2 .$$

Diurnal Flow

It is not necessary to make an adjustment in the design hydraulic loading for normal diurnal wastewater flow patterns when designing for BOD

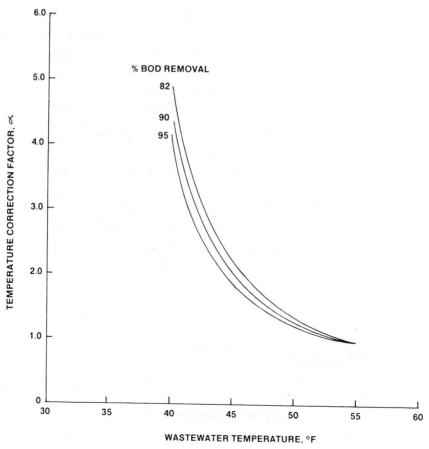

FIGURE 4-3. Temperature correction for BOD removal.

removal. Lower treatment levels occurring at flows above average are balanced by the higher treatment levels occurring at flows below average. The net effect of a cyclic flow pattern is a degree of treatment equivalent to that obtained at a uniform flow equal to average daily flow as long as peak-to-average flow ratio is not excessive (≤3.0).

Suspended Solids Removal

A rotating contactor plant designed to achieve a specific effluent BOD value will also have an effluent suspended solids concentration equal to or lower than the BOD value. Figure 4-4 and Table 4-3 contain data correlating effluent suspended solids with effluent BOD. The majority of the data points in Figure 4-4 fall near or below the 45° line indicating that effluent solids will be equal to or less than effluent BOD.

NITROGEN CONTROL

Nitrogen control has become a wastewater treatment requirement in many parts of the U.S. In many eastern and midwestern states, nitrogen control requirements consist of ammonia nitrogen removal to reduce its oxygen demand on receiving waters. In some cases, ammonia nitrogen removal is required to keep free ammonia concentrations in receiving waters below levels considered potentially toxic to aquatic life. In still other cases, ammonia nitrogen removal is desired to reduce chlorine requirements and chlorine contact times for effective disinfection. Ammonia removal and reduced chlorine dosage also help avoid forming chloramines and chlorinated organic compounds which are toxic to aquatic life. Ammonia nitrogen removal can be achieved through biological oxidation of ammonia nitrogen to nitrate nitrogen. While this does not remove ammonia nitrogen from the wastewater, it does convert it to a form which meets each of the requirements just stated.

In some areas of the U.S., removal of all forms of nitrogen from the wastewater is required. In areas such as Long Island, New York, where wastewaters are discharged to ground aquifers to become part of the water supply, it is necessary to limit the nitrogen content of the effluent to prevent an accumulation of nitrates and exceed public health standards for drinking water. In other areas, such as Florida and some mid-Atlantic states, the total nitrogen content of wastewater discharges is being limited as a means of controlling eutrophication and avoiding undesired development of algae in receiving waters. To meet total nitrogen requirements for all of these cases, it is first necessary to nitrify or aerobically oxidize ammonia nitrogen to nitrate and then anaerobically denitrify or convert the nitrates to nitrogen gas. Denitrification is an anaerobic process and usually requires the addition of organic carbon to the wastewater to provide a source of energy for the denitrifying bacteria. They utilize the oxygen contained in the nitrate ion in their metabolic processes. Methanol is most often considered as the source of carbon, because it is relatively inexpensive, easy to handle, and most important, results in relatively little biological sludge generation.

In some isolated instances, land disposal of wastewater has become an issue. In these cases, nitrification and denitrification treatment is an alternative to consider, because it will produce an effluent comparable in quality to land disposal of wastewater. A significant difference, however, is that with nitrification/denitrification treatment, final effluent quality is more closely controlled, and there is much less danger that ground water supplies could become contaminated. The rotating biological contactor process has been tested for application to both nitrification and denitrification treatment requirements.

Nitrification

Previous testing of the process described in Chapter 3 has shown that it can easily achieve high degrees of both BOD removal and nitrification in a single step of treatment. Microorganisms which primarily remove carbonaceous materials predominate in the early stages of treatment, while nitrifying bacteria develop in the latter stages. All that is necessary for development of a nitrifying culture is to reduce the BOD concentration to the point where nitrifying bacteria can effectively compete with the more rapidly growing heterotrophic organisms which oxidize the organic carbon present.

Since the completion of the initial studies mentioned above, additional pilot plant tests have been conducted at other municipalities around the U.S.[4] These tests have used pilot plants similar to that shown in Figure 4-5. The unit contains 18-in.-diameter polyethylene media mounted on a single shaft and divided into 4 equal-sized stages along the length of the shaft. The stages are

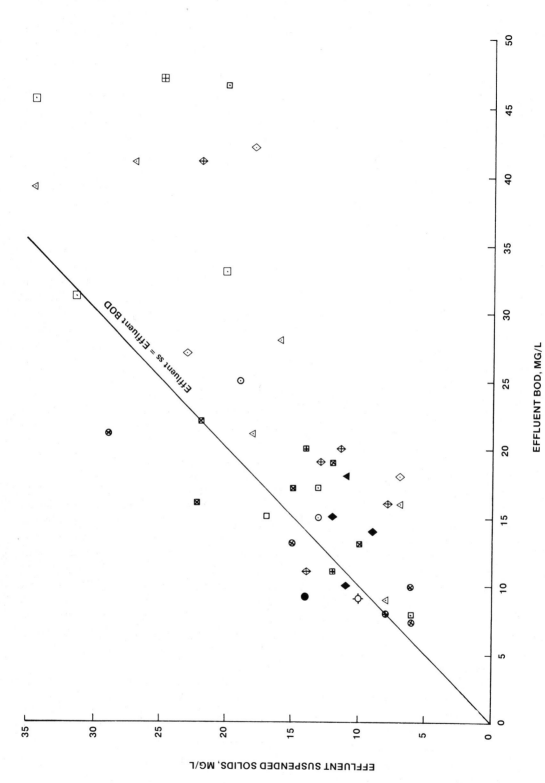

FIGURE 4-4. Municipal wastewater treatment-effluent suspended solids.

TABLE 4-3

Municipal Wastewater Treatment-effluent Suspended Solids

0.5-m diameter pilot plants	Full-scale plants and 3.2-m diameter pilot plants
△ —Pewaukee, Wis.	□ —Philadelphia, Pa.
□ —Spencer, Iowa	◇ —Gladstone, Mich.
◇ —Bemidji, Minn.	○ —Rehrersburg, Pa.
○ —Spring City, Pa.	● —Omaha, Neb.
○ —Washington, Pa.	▲ —Niagara, Wis.
□ —Philadelphia, Pa.	○ —Pewaukee, Wis.
◆ —Shelbyville, Ind.	

FIGURE 4-5.   Pilot plant for nitrification studies.

separated by cross-tank bulkheads, and flow passes from stage to stage through submerged orifices in the bulkheads. The unit was normally tested over a range of flows of 200 to 1,000 gpd. Pilot plant testing at the various communities was supervised jointly by the personnel from the municipalities and by the municipalities' consulting engineers.

Test data from these locations are shown in Figure 4-6. In Figure 4-6, the degree of ammonia nitrogen removal is related to the hydraulic loading on the rotating media as gallons per day of wastewater flow per square foot of available surface area covered with biological growth. The test data are divided into several BOD concentration ranges to show the influence of inlet BOD strength on the degree of nitrification. For a primary effluent, BOD concentration of 150 mg/l, 90% nitrification can be achieved at a loading rate of approximately 1.5 gpd/ft², and 95% nitrifica-

tion can be achieved at a loading rate of 1.25 gpd/ft². For lower influent BOD concentrations, higher hydraulic loadings can be used, because nitrifying bacteria will begin to appear in the earlier stages of media. Conversely, for high influent BOD concentrations, lower hydraulic loadings must be used, because nitrifying bacteria will first appear in the latter stages. The data show that very high degrees of nitrification can be achieved for any influent BOD concentration simply by selecting an appropriate design hydraulic rate.

The data in Figure 4-6 were used to arrive at the design criteria shown in Figure 4-7. This is essentially the same correlation with two changes. One is that there is a maximum ammonia nitrogen concentration for which the data are considered valid. This concentration is generally in the range of $\frac{1}{5}$ to $\frac{1}{10}$ the BOD concentration. It will be

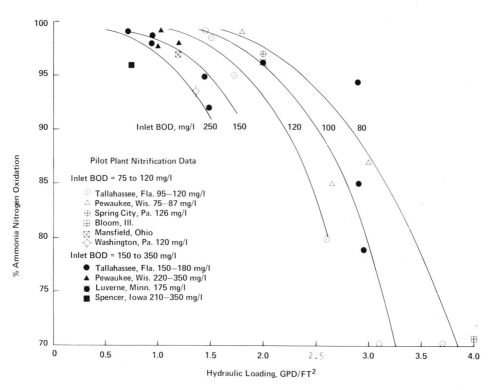

FIGURE 4-6. Pilot plant nitrification data.

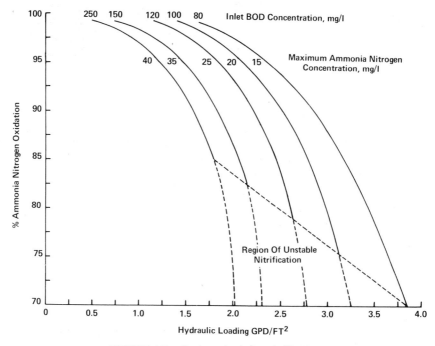

FIGURE 4-7. Design criteria for nitrification.

unusual if the ammonia nitrogen present for a particular application exceeds the limits shown in Figure 4-7. When it does, however, use the design curve which is valid for that ammonia nitrogen concentration.

Peak Flow Correction

The second change from Figure 4-6 is the identification of a region of unstable nitrification, that is, a region of hydraulic loading where either a slight change in the hydraulic loading or the influent BOD strength could result in a significant loss of nitrification efficiency. It is recommended that the process not be designed to operate in that region. Also, it is advisable to avoid entering that region of hydraulic loading during daily peak flow conditions. During daily flow variations, ammonia nitrogen removal efficiency can be estimated by moving back and forth along the lines shown in Figure 4-7. If, however, during peak flow conditions, the region of unstable nitrification is entered for a sufficient period of time (more than 3 hr), the nitrifying population could be displaced by more rapidly growing heterotrophic organisms. Therefore, during low flow conditions, nitrification efficiency would n ɔt be restored, because the slow-growing nitrifying population could not redevelop quickly enough.

Figure 3-40 indicates how this could happen. It shows the effect of the effluent BOD produced on the degree of nitrification achieved. For effluent BOD concentrations of 30 mg/l and above, there is relatively little nitrification occurring in the system. However, once wastewater BOD is reduced to less than 30 mg/l, the nitrifying bacteria can grow rapidly enough to predominate in the latter stages of treatment. What is necessary, then, is to avoid developing high BOD concentrations in these latter stages of treatment under peak flow conditions. According to Figure 3-40, this could be done effectively if the effluent BOD from the process did not exceed approximately 18 mg/l under peak flow conditions. To achieve this, it is necessary to design the process for a BOD removal efficiency under peak daily flow conditions, so that 18 mg/l BOD is not exceeded in the effluent. This criterion, along with Figure 4-1, was used to develop allowable peak-to-average diurnal flow ratios as a function of degree of nitrification and influent BOD strength as shown in Figure 4-8. This design criterion, although somewhat arbitrary and based on steady state operation, is thought to be conservative since it is based on a relatively low effluent BOD level. Keep in mind that the 18 mg/l includes the 5-day nitrogenous oxygen demand and the BOD associated with unsettled biological solids so that soluble carbonaceous BOD will be less than 10 mg/l.

For a primary effluent BOD strength of 150 mg/l and a requirement for 90% nitrification, Figure 4-8 indicates that the peak-to-average flow must not exceed 1.6 to avoid producing an effluent of greater than 18 mg/l BOD. At a higher level of nitrification (95% ammonia removal), the allowable peak-to-average flow ratio increases to 1.95. For higher influent BOD strengths, the allowable peak-to-average ratio decreases for all levels of nitrification, and for lower influent strengths, it increases for all degrees of nitrification.

For wastewater treatment applications, where the allowable peak-to-average flow ratio is exceeded, there are two alternatives to satisfy the requirement. One is to increase the average design flow rate (or decrease the design loading rate), so that the peak-to-average flow ratio meets the requirement. This will result in a larger amount of rotating surface area for the application. The other alternative is to install flow equalization of sufficient capacity to meet the allowable peak-to-average flow ratio. The choice between the two alternatives will depend upon whether it is more cost effective to install more rotating surface area or to install flow equalization. For smaller plants, it is likely that more rotating surface area will be more economical, while for larger plants, it is likely that flow equalization will be more economical, since it will also reduce equipment requirements for pretreatment, final clarification, and any tertiary treatment steps that may follow.

Changes in BOD and ammonia nitrogen concentrations can also affect the nitrification efficiency of a biological treatment process. The data in Figure 4-6 for the rotating contactor process were collected at municipalities where normal diurnal concentration variations occurred so that this effect is incorporated into the design relationships.

Temperature Correction

Wastewater temperatures below 55°F will affect nitrification efficiency with the rotating biological contactor process just as it will with all types of biological treatment. Several pilot plants like that shown in Figure 4-5 were operating at

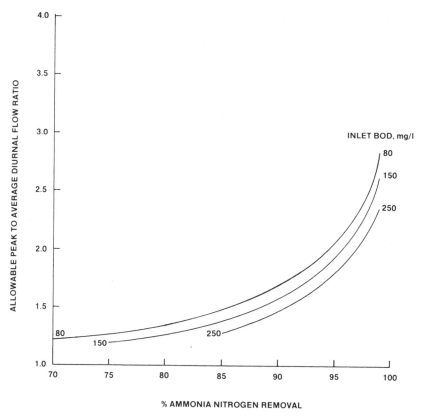

FIGURE 4-8.   Design criteria for flow variations.

Pewaukee, Wisconsin during the winters of 1972 to 1974. The test data generated from them are shown in Figure 4-9. To correlate this data, lines of equal hydraulic loading have been drawn through the various points for decreasing wastewater temperatures. The lines were drawn by starting at 55°F with a point established from the hydraulic loading relationships of Figure 4-7 and then continuing through similar hydraulic loading points for lower wastewater temperatures. The data show that for low hydraulic loadings and high degrees of nitrification, there is relatively little loss of nitrification efficiency as wastewater temperature decreases. For higher hydraulic loadings and lower degrees of nitrification, ammonia nitrogen removal efficiency decreases more rapidly as the wastewater temperature decreases. This phenomenon has also been found to a lesser degree for BOD removal and can perhaps best be explained in terms of sludge age. At high loading rates where the culture is in a relatively high state of activity, or short sludge age, temperature has a significant effect on ammonia nitrogen removal. At low loading rates, the culture is relatively inactive or

has a long sludge age and is operating under ammonia limiting conditions. Under these conditions, temperature has relatively little effect on nitrification efficiency.

Temperature correction factors were developed from the family of hydraulic loading lines in Figure 4-9. This was done by calculating the ratio of the hydraulic loadings for a specific degree of nitrification and for the different wastewater temperatures. These ratios or temperature correction factors were then plotted as a function of degree of nitrification and wastewater temperature in Figure 4-10. These temperature correction factors are to be used to reduce the design hydraulic loading rate determined from Figure 4-7 for any wastewater temperature lower than 55°F.

When designing a rotating contactor process for simultaneous BOD and ammonia nitrogen removal, the hydraulic loading rate for each must be determined separately using Figures 4-1 and 4-7. The loading rate for nitrification must then be corrected for peak flow conditions if necessary. Both loading rates must then be separately corrected for temperature, if necessary, using

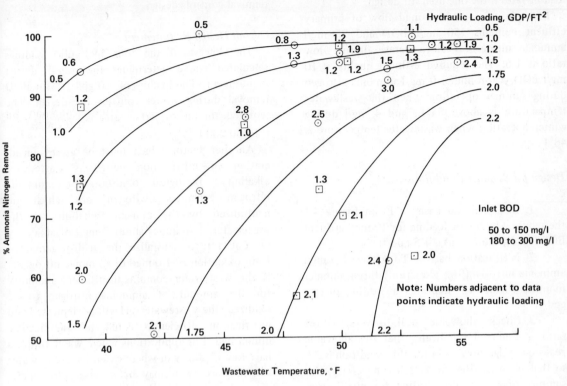

FIGURE 4-9.   Effect of wastewater temperature on nitrification.

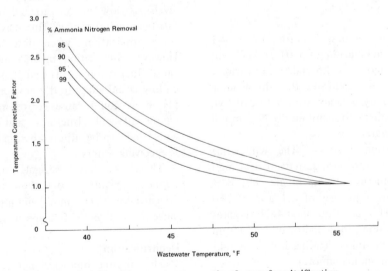

FIGURE 4-10.   Temperature correction factors for nitrification.

Figures 4-3 and 4-10. The lower of the two final loading rates is the one used for design.

Example #3 — A 1.0 mgd flow of primary effluent contains 150 mg/l BOD and 25 mg/l ammonia nitrogen. Peak-to-average diurnal flow ratio is 1.6. The required effluent quality is 10 mg/l BOD year-round, 1.0 mg/l ammonia nitrogen during summer operation when the wastewater temperature is above 55°F, and 4 mg/l during winter operation when wastewater temperature is 48°F.

Design for Summer Conditions

1. BOD removal basis — From Figure 4-1 the required hydraulic loading for 10 mg/l effluent BOD (93.3% reduction) is 1.5 gpd/ft^2.

2. Nitrification basis — To achieve 1.0 mg/l ammonia nitrogen (96% reduction) during summer months, Figure 4-7 indicates a loading of 1.3 gpd/ft^2.

3. Check allowable peak-to-average flow ratio — Figure 4-8 indicates that the allowable peak-to-average flow ratio for this condition is 2.0 so that no peak flow correction is necessary. For summer operation, the loading for nitrification controls design.

Design for Winter Conditions

1. BOD removal basis — From Figure 4-1 the required hydraulic loading for 10 mg/l effluent BOD (93.3% reduction) is 1.5 gpd/ft^2. For winter operation, Figure 4-3 indicates that this loading must be reduced by a factor of 1.5 to a design loading of 1.0 gpd/ft^2 to maintain 93.3% removal with 48°F wastewater.

2. Nitrification basis — The wintertime effluent ammonia nitrogen requirement of 4 mg/l is just 84% reduction. From Figure 4-7, 84% reduction requires a loading of 2.1 gpd/ft^2. To maintain this level of nitrification at 48°F requires a reduction in loading by a factor of 1.45 (from Figure 4-10). The required loading at 48°F is 1.45 gpd/ft^2 for 84% ammonia removal.

3. Check allowable peak-to-average flow ratio — At this relatively low degree of nitrification, Figure 4-8 indicates that the allowable peak-to-average flow ratio is just 1.35. The loading rate must be reduced by this ratio ($\frac{1.35}{1.6} \times 1.45 =$ 1.22) to assure that nitrifying organisms will not be displaced during peak flow conditions. For winter operation, the loading rate for BOD removal controls design.

Overall Design Requirement

The lowest of the four hydraulic loadings calculated above determines the overall design requirement. The loading of 1.0 gpd/ft^2 for BOD removal during winter controls the design. The required amount of surface area is $\frac{1,000,000 \text{ gpd}}{1.0 \text{ gpd/ft}^2}$ = 1,000,000 ft^2.

Another factor which must be given consideration in nitrification design is wastewater alkalinity. Biological oxidation of ammonia nitrogen generates hydrogen ions which are neutralized by the available alkalinity in the wastewater. It requires about 7 mg/l of alkalinity (as $CaCO_3$) to neutralize the acidity generated from oxidation of 1.0 mg/l of ammonia nitrogen.[5] If the wastewater contains insufficient alkalinity for the amount of ammonia nitrogen to be oxidized, the wastewater pH will be depressed and nitrification efficiency could be significantly diminished. For applications where water supplies have low alkalinity or where extensive use of water softening exists, it may be necessary to supplement the alkalinity through chemical addition. Haug and McCarty[5] have shown that fixed nitrifying cultures can adapt to consistent pH levels as low as 5.5 to 6.0 without loss of nitrification efficiency. This may be possible for some applications where low alkalinity exists. However, fluctuation in alkalinity levels would cause fluctuation in pH and it is unlikely that a culture could adapt to this environment. Also, low pH levels could cause corrosion problems with equipment and structures, and in many cases would have to be adjusted before being discharged to receiving waters.

There are now several full-scale rotating contactor plants in operation which have been designed for BOD removal and nitrification. These range in size from 20,000 gpd to 1.2 mgd capacity.

Denitrification

The rotating biological contactor process has also been investigated for denitrification. Denitrification with the process is achieved by completely submerging the rotating media and with the addition of an appropriate source of organic carbon, anaerobic denitrifying bacteria develop on the surfaces. Figure 4-11 is a process flow diagram for this application.

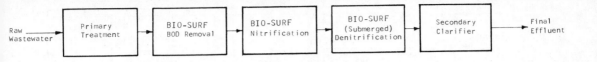

FIGURE 4-11.   Denitrification schematic.

Figure 4-12 shows a small treatment plant utilizing the process for all three steps of biological treatment. After primary treatment, the wastewater is distributed to two parallel operating shafts of media to provide BOD removal and nitrification. These two effluents then combine and flow to a third completely submerged shaft with methanol addition to provide the biological denitrification. This effluent then goes to a final clarifier. One significant advantage for the rotating contactor process over a suspended growth system in this application is that all three steps of biological treatment are provided and only a single, final clarifier is necessary. No sludge recirculation streams are necessary. Staged operation of the denitrification step achieves better utilization of the supplemental carbon source to reduce operating costs. The supplemental carbon source is a potential contaminant of the effluent, which must be carefully controlled to avoid exceeding maximum allowed effluent BOD values. This is more easily achieved with staged operation.

An advantage the rotating contactor has over other types of fixed growth reactors for denitrification is that it never has to be backwashed to remove accumulated excess biomass. Rotation of the submerged media exerts a shear sufficient to continuously strip all excess biomass from the media. This eliminates the backwashing equipment and controls and also eliminates the need for duplicate process equipment to permit continuous operation during backwashing operations.

For a large plant, the schematic shown in Figure 4-13 would be used. Here, wastewater would enter and flow perpendicular to the shafts of rotating media for BOD removal and nitrification and then proceed directly to submerged shafts of rotating media for denitrification. Drive assemblies are mounted on top of the tank wall and are essentially identical to those used for partially submerged media. Small gaps are placed along the length of the media assembly to promote internal mixing since gravity no longer assists mixing efficiency. A floating plastic cover can be placed on the water surface to restrict oxygen absorption and reduce requirements for the supplemental source of organic carbon.

Pilot plant tests have been conducted at several locations around the U. S. utilizing a test unit like that shown in Figure 4-14. It contains polyethylene media 2 ft in-diameter mounted on a horizontal shaft with the media divided into 4 equal-sized stages and with cross-tank bulkheads separating the media into 4 successive stages of treatment. Preliminary results obtained thus far indicate that for nitrate nitrogen concentrations up to 10 mg/l, 90 to 95% denitrification can be obtained at a hydraulic loading of 5 gpd/ft^2. For nitrate nitrogen concentrations up to 25 mg/l, 90 to 95% denitrification occurs at a loading of 3 gpd/ft^2.

Using 90% denitrification at 3 gpd/ft^2 as a basis, and assuming first order behavior, Figure 4-15 also shows loadings required for lower degrees of denitrification. For some applications, only moderate degrees of denitrification will be necessary to meet total nitrogen requirements. Figure 4-15 is preliminary and is meant only for estimating purposes. While it represents the best available information at this time, it should not be used for final design. It will be confirmed through continued testing.

Because of the need for a high degree of nitrification preceding denitrification, a design for nitrification is the first step in the denitrification design procedures. An effluent level of 1.0 mg/l ammonia nitrogen is consistently achievable with the rotating contactor process. This effluent ammonia nitrogen concentration is the basis for several calculations which follow.

The process surface area requirements preceding denitrification are determined using Figures 4-1 and 4-7 as described in Example 3.

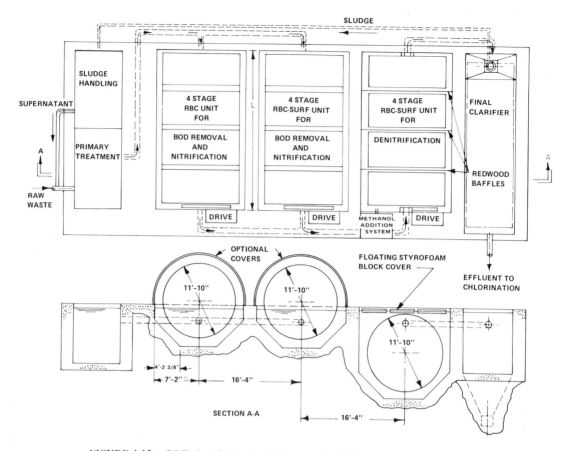

FIGURE 4-12.   RBC plant design for BOD removal, nitrification and denitrification.

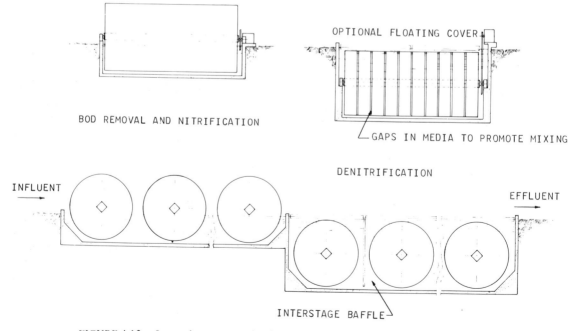

FIGURE 4-13.   Large plant construction for BOD removal, nitrification, and denitrification.

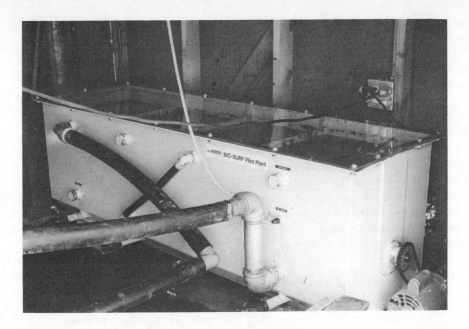

FIGURE 4-14.   Denitrification pilot plant.

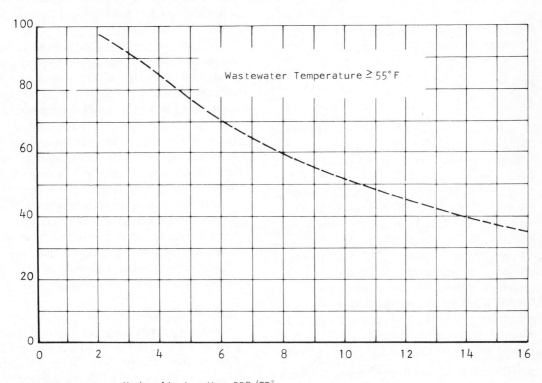

Hydraulic Loading GPD/FT²

FIGURE 4-15.   Tentative design correlation for denitrification.

A rotating contactor treatment plant producing an effluent with 1.0 mg/l ammonia nitrogen, will have 2 to 3 mg/l of total Kjeldahl nitrogen (including ammonia) in the effluent (see Chapter 8). The organic nitrogen in the raw wastewater is removed from the system principally through cell synthesis to an effluent level of 0.5 to 2 mg/l. Approximately 90% of ammonia nitrogen removed is converted to nitrate nitrogen (see Chapter 8). The balance is accounted for in cell synthesis and possibily some denitrification. To be conservative, a complete conversion of ammonia to nitrate should be used for denitrification design purposes.

Example #4 — A wastewater flow of 1.0 mgd has the following characteristics after primary treatment:

	mg/l
BOD	150
Kjeldahl nitrogen	40
Ammonia nitrogen	25
Nitrate nitrogen	0
Total nitrogen	40

It is desired to produce a final effluent of the following characteristics:

	mg/l
BOD	10
Total nitrogen	3

A consistently achievable effluent ammonia nitrogen concentration in a rotating contactor effluent is 1.0 mg/l. The required surface area for BOD removal and nitrification is indicated in Example #3. The amount of Kjeldahl nitrogen in the effluent from the BOD removal and nitrification step at 1.0 mg/l effluent ammonia nitrogen is 2.0 mg/l. For a final effluent of 3 mg/l total nitrogen, there can then be 1.0 mg/l of nitrate nitrogen remaining after denitrification. The amount of nitrate nitrogen in the effluent from the BOD removal and nitrification steps with the assumption of 100% conversion is 25 mg/l. The required degree of denitrification is then 96%. From Figure 4-15, this requires $\frac{1,000,000 \text{ gpd}}{2 \text{ gpd/ft}^2} =$ 500,000 sq ft of completely submerged rotating contactor surface area.

Wastewater temperature will affect the rate of denitrification just as it affects BOD removal and nitrification. Until more specific criteria are developed for temperature effects on denitrification, the temperature correction factor for BOD removal in Figure 4-3 can be used. Because these two reactions are each performed by heterotrophic organisms (those that need organic carbon), the temperature effects are likely to be similar.

OTHER DESIGN CONSIDERATIONS

Secondary Clarifiers

Secondary clarifiers following the rotating contactor process can be designed in accordance with accepted standards (such as *Ten States Standards* and Water Pollution Control Federation (WPCF) Manual No. 8, *Sewage Treatment Plant Design*). For most applications, clarifier overflow rates of 800 gpd/ft^2 are recommended.

When effluent suspended solids levels as low as 10 mg/l are required, somewhat lower overflow rates are recommended, in the range of 400 to 600 gpd/ft^2, depending upon expected diurnal flow patterns. For daily peak to average flow ratios of greater than 2:1, an overflow rate of 400 gpd/ft^2 is recommended for required effluent suspended solids levels of 10 mg/l. For lower flow ratios, 600 gpd/ft^2 is recommended.

Because rotating contactor effluent will generally contain 100 to 150 mg/l of suspended solids, solids loading on the clarifier is quite low. Solids separation occurs by discrete particle settling with no hindered settling or solids compression. Therefore, surface overflow rate is the only important factor in clarifier design. Because there is no significant sludge blanket, the clarifier can be relatively shallow, nominally 7 ft deep, although shorter depths will often be adequate. Either circular or rectangular clarifiers can be used. Simple mechanical solids collectors are recommended over the suction or siphon type because rapid recirculation of solids is not used and because suction collector systems are expected to unnecessarily dilute the waste sludge.

Sludge Handling and Disposal

Sludge production by the rotating contactor process is shown in Figure 4-16. Sludge solids are 80% volatile (Table 3-3). In a conventional secondary clarifier, settled sludge thickens to 3 to 4% solids. Sludge hopper walls should have a minimum slope of 2:1. If sludge is drawn off intermittently in small volumes, the high solids concentration can be maintained. If it is drawn off frequently at high rates, it will be diluted to 1 to 2% solids. If secondary sludge is recycled to the

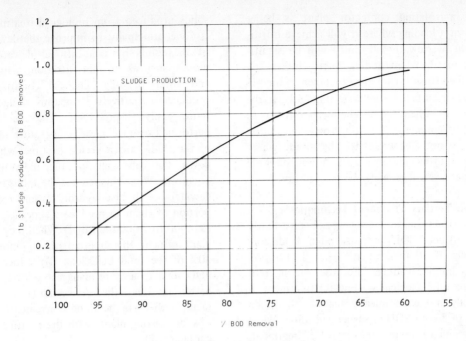

FIGURE 4-16. Sludge production.

primary clarifier, it will thicken with primary sludge to 4 to 6% solids. This recycling procedure is generally recommended for maintaining high sludge solids concentration. Further thickening prior to sludge treatment and disposal is not necessary.

Aerobic digestion or heat treatment of secondary sludge is often utilized for waste sludge disposal, however, the high solids content of rotating contactor process sludge warrants consideration of anaerobic digestion as a means of sludge treatment. There may also be cases where anaerobic digestion gases could run engine-generators to supply power to operate the treatment plant. The low power requirements of the process can, in some cases, make this an attractive alternative.

Primary Treatment Requirements

Primary treatment for the rotating contactor process can consist of a conventional primary clarifier, or a combined primary clarifier and anaerobic sludge digester often called an Imhoff Tank. These are to be designed in accordance with accepted criteria as defined in standard references such as *Ten States Standards* or WPCF Manual No. 8 *Sewage Treatment Plant Design*.

Fine screening equipment recently developed by several manufacturers preceded or followed by grit removal (when necessary), is also an acceptable means of providing primary treatment. Bar screening or comminution, however, are not suitable as the sole means of primary treatment because of potential maintenance problems from accumulation of grit and other dense solids in the rotating contactor tankage.

Chlorination Requirements

Chlorine dosage required for disinfection of rotating contactor effluent is equal to that of any biological process providing the same degree of treatment. The chlorine dosage of 8 mg/l recommended by *Ten States Standards* for activated sludge process effluents can be used for rotating contactor plants providing 90% BOD removal. For plants designed for nitrification, the chlorine dosage can be reduced to 3 mg/l or less.

Phosphate Removal

Phosphate removal by the rotating biological contactor process is similar to other biological treatment processes, i.e., approximately 20% removal. Additional phosphate removal can be obtained by adding chemicals to the primary clarifier or to the rotating contactor tankage. Phosphate removal in excess of 90% has been achieved with the addition of ferric chloride or alum directly to the rotating contactor tankage.[6,7]

When a rotating contactor plant is being operated at a loading where it will achieve nitrification, the wastewater alkalinity will be significantly reduced. When adding ferric chloride or alum to the effluent, care must be taken so the pH is not depressed below discharge standards. In the same manner, when adding chemicals directly to the rotating contactors, care must be taken so that the pH is not decreased to the point where nitrification efficiency is reduced. A consistent pH of 6.5 or above will generally be satisfactory.[5]

Applicable Tertiary Treatment Techniques

The rotating contactor process can consistently produce effluent BOD and suspended solids concentrations of 10 mg/l when operated at appropriate hydraulic loadings. Because of inherent limitations on biological wastewater treatment, it is generally not economically feasible to obtain effluents of 5 mg/l BOD or suspended solids. When effluent standards require this degree of treatment, it will be necessary to utilize a tertiary or advanced treatment step following the rotating contactor process. Applicable tertiary treatment techniques are any of those already being used to follow other biological treatment systems and include microscreening, pressure or gravity filtration, adsorption on activated carbon, and lagoons.

Criteria for lagoon design to follow a rotating contactor process are identical to those for any biological treatment process providing the same degree of treatment.

Pilot plant testing of pressure filtration was conducted on effluent from the rotating contactor unit shown in Figure 3-1. The filter was 18 in. in diameter and contained a 1-ft depth of 1.0 mm sand. It was operated at 15 gpm/ft^2 and backwashed at a pressure drop of 15 psi. Results of the tests are presented in Figures 4-17 and 18. They show 40 to 50% reduction of BOD and suspended solids and effluent concentrations as low as 3 mg/l BOD and suspended solids.

Pilot plant tests by Torpey[8] have shown activated carbon to be an effective means of providing tertiary treatment. Effluents of 10 mg/l total organic carbon can be reduced to 2 mg/l with activated carbon treatment of rotating contactor effluents.

Sludge Treatment Supernatant

During the past several years, there has been a growing interest in thermal conditioning of sludge solids produced from biological treatment, which is done principally to improve the dewaterability of the solids by conventional dewatering techniques, such as vacuum filtration, centrifugation, and filter pressing. Thermal conditioning also produces a relatively innocuous sludge which is more easily disposed of. The conditioning is performed by raising the temperature and pressure of the sludge for a period of time, which breaks down the cell walls of the individual microorganisms. While this renders the solids more easily dewaterable, it also hydrolyzes and redissolves a portion of them to produce a supernatant as high as 5,000 to 7,000 mg/l BOD. When recycled to the plant influent, this supernatant can represent 15 to 30% of the total BOD load. To reduce this extra BOD load to a more manageable level, it is often better to treat the supernatant in a separate facility; this can be done conveniently in either new or existing plants with the rotating biological contactor.

Pilot plant tests conducted at several locations indicate that loading rates of less than 1.0 gpd/ft^2

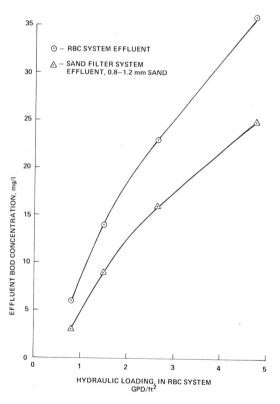

FIGURE 4-17. Tertiary BOD reduction by sand filtration of RBC effluents.

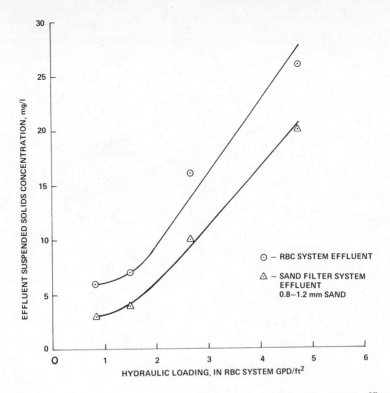

FIGURE 4-18. Tertiary suspended solids reduction by sand filtration of RBC effluents.

are necessary to treat the supernatant to a BOD concentration equivalent to domestic sewage. For supernatants as concentrated as 7,000 mg/l, loading rates as low as 0.2 gpd/ft² may be required. The exact loading rate necessary may differ from plant to plant, so that pilot plant studies are normally recommended.

Anaerobic digester supernatant is similar in nature to supernatant from thermal sludge conditioning and can also be treated by a separate rotating biological contactor system before being recycled. A large, full-scale municipal plant is under construction in 1975 in which two rotating contactor shafts are to be used for partial treatment of centrate from centrifuge dewatering of anaerobically digested sludge.

Lagoons used for long-term storage of sludge can develop a supernatant which, while relatively low in BOD content, can have very high concentrations of ammonia nitrogen. If discharged to a receiving water, the ammonia would exert a significant oxygen demand. To avoid this, the supernatant can be biologically treated to oxidize the ammonia nitrogen. This has been done by Lue-Hing et al.[9] using a rotating contactor pilot plant. They found that a lagoon supernatant containing about 100 mg/l BOD and 800 mg/l

ammonia nitrogen could be treated to less than 2 mg/l ammonia nitrogen at a rotating contactor loading rate of 0.1 gpd/ft² and a peripheral media velocity of 15 ft/min.

Biological nitrification of high ammonia wastes is sometimes difficult to accomplish because of an apparent inhibition from an accumulation of nitrous acid. This did not appear to be a problem in the studies of Lue-Hing, because effluent nitrite concentrations were less than 1.0 mg/l. However, interstage samples showed nitrite concentrations as high as 100 mg/l in the intermediate stages along with reduced ammonia oxidation rates.

PACKAGE PLANTS

The rotating contactor process is well suited for the treatment of low wastewater flows where process simplicity and low maintenance are critical requirements. Low power consumption, stable operation under large flow fluctuations, and ease of sludge handling also contribute to the attractiveness of the process for this application.

A form of process equipment for this application is shown in Figure 4-19. It consists of a feed chamber and rotating bucket feed mechanism, a multistage media assembly, a secondary clarifier

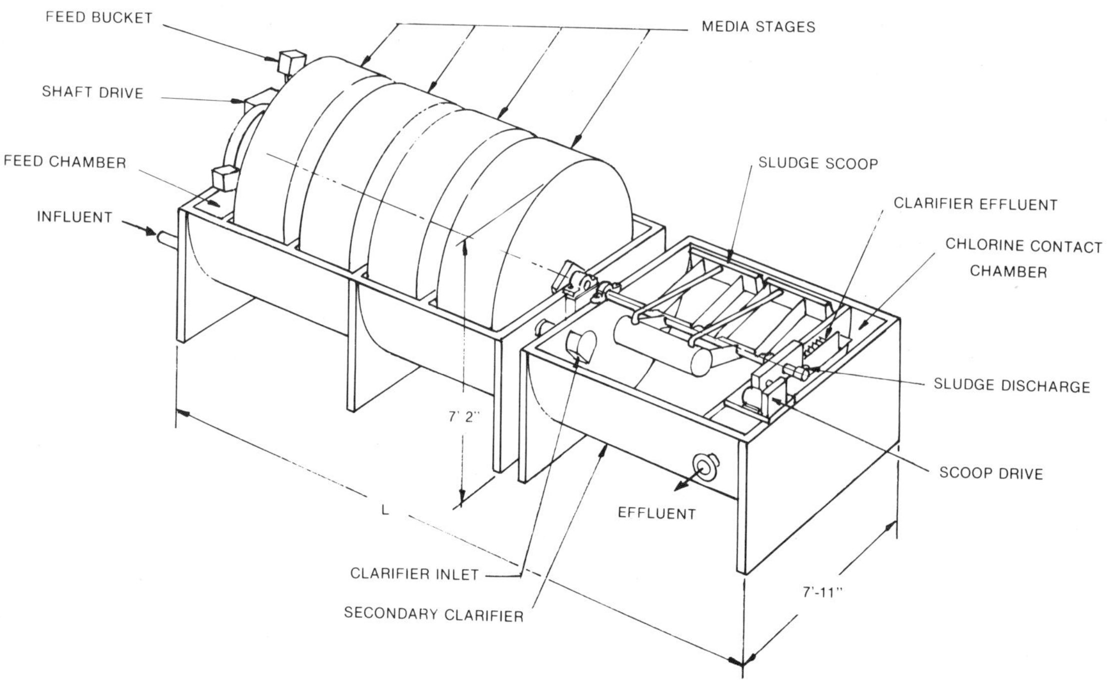
FEED BUCKET
MEDIA STAGES
SHAFT DRIVE
FEED CHAMBER
SLUDGE SCOOP
INFLUENT
CLARIFIER EFFLUENT
CHLORINE CONTACT
CHAMBER
SLUDGE DISCHARGE
SCOOP DRIVE
7' 2"
L
EFFLUENT
CLARIFIER INLET
SECONDARY CLARIFIER
7'-11"

FIGURE 4-19. Package RBC plant.

with rotating sludge scoop, and a chlorine contact tank. All this equipment is incorporated into semicircular steel tanks. The system is intended to operate in conjunction with primary treatment and sludge disposal facilities.

Aerobic Pretreatment

Several aerobic pretreatment alternatives can be utilized to precede package units. These include primary clarifiers, combined primary clarifier and sludge storage (Imhoff tank), or any of the recently developed fine-screening devices. The use of combined primary clarification and sludge storage is shown in Figure 4-20. Secondary sludge from the package unit is returned to the pretreatment system for storage and digestion. Because package plant applications often experience extreme fluctuations in flow, it is often recommended that an aerated flow equalization step be utilized after pretreatment before entering the treatment unit. This results in a higher treatment capacity and permits a more economical design for biological treatment and the subsequent clarification and disinfection steps.

Septic Tank Pretreatment

For wastewater flows up to 50,000 gpd, primary treatment and sludge handling can be accomplished simply and effectively with a septic tank. A septic tank by itself is a crude means of wastewater treatment. However, in combination with the rotating contactor process, it provides a very simple package plant for secondary treatment.

A prototype package plant like that shown in Figure 4-21 (and Figures 3-1 and 3-2) was tested on septic tank effluent at Pewaukee, Wisconsin in 1971. The test unit operated with a buried 4,000-gal septic tank as shown in Figure 4-21.

Test data collected over a 9-month operating period are shown in Figures 4-22 and 4-23. Comparing these removal efficiencies for BOD and ammonia nitrogen to those from a similar unit treating fresh wastes indicates that approximately 50% more surface area is required to provide equivalent treatment on a septic waste. This is probably due to the necessity to convert the wastewater to aerobic conditions and develop a biological culture which has become acclimated to these conditions.

Two configurations with a septic tank are shown conceptually in Figures 4-24 and 4-25. For each of these configurations, the pretreatment system is constructed of two parts – the septic tank and a flow equalization tank. For low wastewater flows,

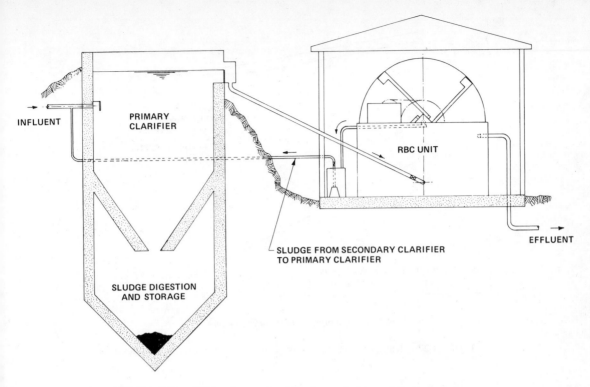

FIGURE 4-20.   RBC package unit with primary clarifier and sludge digestion.

FIGURE 4-21.   Prototype package plant with septic tank.

these two tanks may be contained in a single tank. For higher flows, they can be separate tanks. The septic tank portion of the pretreatment system provides for removal of settleable and floatable materials. It also provides for anaerobic digestion and storage of secondary sludge solids. The flow equalization tank in conjunction with the feed chamber and bucket feed system of the rotating contactor unit provides a relatively uniform flow of wastewater to the process regardless of the pattern of raw wastewater flow. Septic tank construction can be of steel or concrete depending upon cost, soil conditions, and regulatory agency requirements.

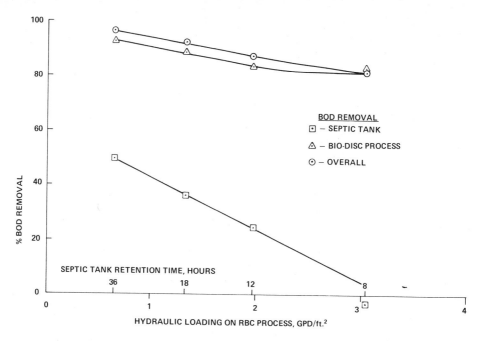

FIGURE 4-22.   RBC process treatment of septic tank effluent for BOD removal.

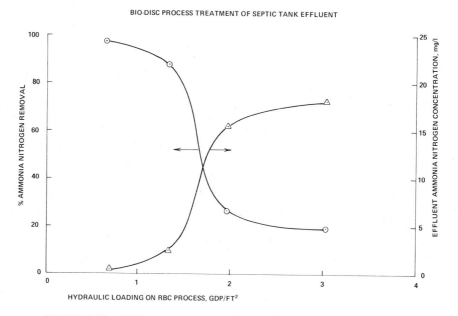

FIGURE 4-23.   RBC process treatment of septic tank effluent for nitrification.

Side-by-side Configuration

In the side-by-side configuration (Figure 4-24), gravity wastewater flow to the unit is utilized. Raw wastewater enters the septic tank for primary treatment and overflows into the flow equalization tank. The flow equalization tank is connected to the lowest point of the feed chamber. At the beginning of a daily flow cycle, the wastewater level in the equalization tank will be at its lowest point. As the flow cycle progresses, the equalization tank will begin to fill and the rising level in the feed chamber will cause the bucket feed mechanism to deliver wastewater at an increasing rate. At the end of the period of flow, wastewater

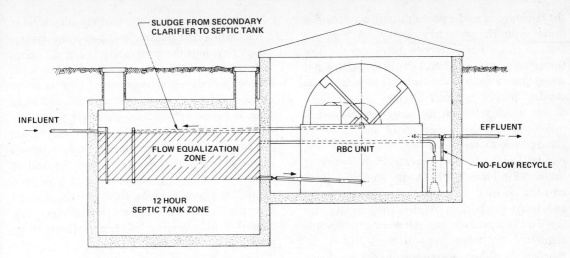

FIGURE 4-24.   RBC package unit with septic tank pretreatment side-by-side configuration.

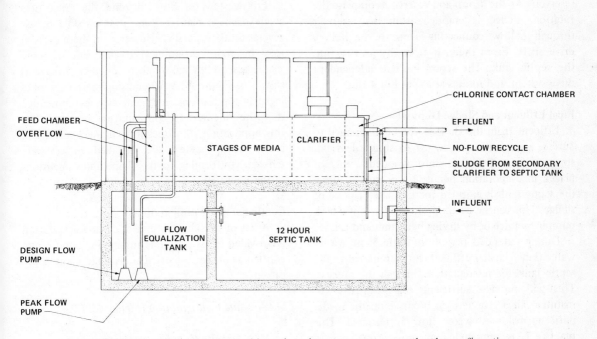

FIGURE 4-25.   RBC package unit with septic tank pretreatment over-and-under configuration.

in the equalization tank and feed chamber will be at their highest levels and the bucket pump will be pumping at its highest rate. During the balance of the daily flow cycle, where there is little or no raw waste flow, the level in the equalization tank will be pumped down by the bucket pumps until the beginning of another flow cycle when it is again at its lowest level. Operating in this manner, the biological treatment process experiences a cyclic wastewater flow pattern where the peak flow is approximately 1.5 times

the average regardless of the pattern of raw wastewater flow. In cases where extreme surges of wastewater flow exceed the capacity of the equalization tank, the excess can overflow from the feed chamber directly into the stages of media and receive full secondary treatment.

Clarifier Operation

Wastewater passes through a submerged orifice in the center of each bulkhead separating individual stages of treatment. Mixed liquor from the last stage of media passes through a submerged orifice into the secondary clarifier. A baffle in front of the orifice uniformly distributes the flow throughout the clarifier. Clarified wastewater passes over a weir in the opposite corner of the clarifier into a baffled chlorine contact tank. Sludge which settles in the secondary clarifier is picked up by the sludge scoop and flows by gravity back to the septic tank for digestion and storage. The sludge scoop operates in the following manner (See Figure 4-19): As the scoop is rotated into the clarifier, it pushes sludge down the side and across the bottom. As the scoop rotates into the 7 o'clock position, the reservoir, to which it is connected by hollow support arms, enters the clarifier. As it does, wastewater pushes the sludge in the scoop through the support arms to fill the reservoir. As the scoop and reservoir complete the rotation, sludge is emptied from the reservoir through hollow connecting arms to the hollow drive shaft. From there, it flows by gravity into the septic tank. The scoop has an independent drive system, and rotates at a speed of 4 rph.

Final Effluent and Sludge Disposal

Effluent from the clarifier can be discharged to surface water, into a new or existing tile field, sprayed onto fields or into wooded areas, or placed in evaporation ponds. From time to time, the septic tank is emptied for ultimate disposal of sludge. This can be accomplished by a septic tank pumping service or by drying beds and land fill.

During extended periods of little or no wastewater flow, clarifier effluent can be pumped to the septic tank for recirculation through the process. This will provide sufficient organic matter to maintain an active aerobic biomass on the media until normal wastewater flow is resumed. This practice is recommended in applications such as recreational areas where wastewater treatment is needed only several days each week.

Over-and-under Configuration

The over-and-under configuration with a septic tank shown in Figure 4-25 operates in a similar fashion to that described for the side-by-side configuration. The major difference is the manner in which the wastewater is fed to the rotating contactor unit. A submersible pump of somewhat greater capacity than average design flow is placed in the flow equalization tank. It pumps at a constant rate into the feed chamber. An adjustable

overflow connection on the feed chamber is set at a level where the buckets will pump the average design flow into the media sections and the excess will overflow back into the equalization tank. During periods when wastewater flow is above average, the equalization tank will fill, and when flow is below average, the tank will empty. In cases where extreme surges in flow could exceed the capacity of the equalization tank, a second submersible pump is placed in the equalization tank. This pump is activated by a high level control and pumps excess flow directly into the first stage of media. In the over-and-under configuration, effluent recycle (when used) returns by gravity to the septic tank.

Media Assembly in Concrete Tankage

For wastewater flows beyond the range of a single package plant system, it is possible to use several units in parallel. However, in these cases, it may be more economical to use one or more media assemblies mounted in concrete tankage as shown in Figure 4-26. When using shaft assemblies in such applications, it is generally recommended that the side-by-side configuration be used. For this application, the bucket feed mechanism and sludge scoop system are installed in concrete tankage conforming to the shape and dimensions of the steel tankage.

Comparison of Configurations

A list of the key features of each configuration is provided to assist in determining the appropriate configuration for a particular package plant application.

Side-by-side Configuration (Figure 4-24)

   · Gravity flow of wastewater.
   · Insulation against cold weather.
   · Can be completely underground.
   · Requires construction of underground chamber to accept package unit. (This duplicates tankage and increases cost.)
   · Requires only a very simple cover.
   · Requires pumping for effluent recycle, when desired.
   · Minimum slope of ¼ in./ft must be maintained on sludge line.
   · Maximum capacity of flow equalization tank limited by available wastewater depth in feed chamber.

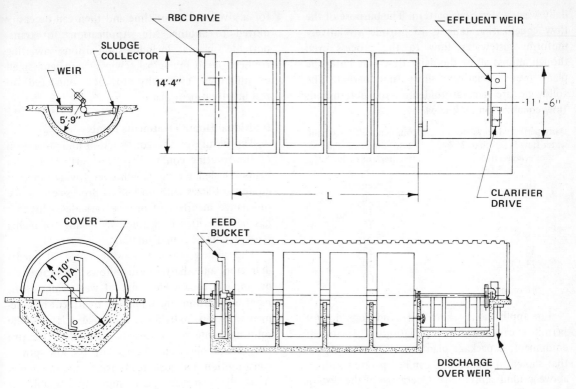

FIGURE 4-26.   RBC package unit assembly for concrete tankage.

Over-and-under Configuration (Figure 4-25)

•   Requires pumping of wastewater into the package unit.
•   Requires construction of a building for the package unit.
•   Gravity return of secondary sludge and effluent recycle (when desired) to septic tank.
•   Full capacity of flow equalization tank can be utilized.

Side-by-side with Media Assembly in Concrete Tankage (Figure 4-26)

•   Gravity flow of wastewater and sludge.
•   Insulation against cold weather.
•   Minimum slope must be maintained on sludge line.
•   Construction can be completely underground.
•   A single concrete tank required for rotating shaft assembly.
•   Requires only a simple cover.
•   Requires pumping for effluent recycle when desired.

•   Maximum capacity of flow equalization limited by available wastewater depth in feed chamber of concrete tankage.

Design Criteria

Design criteria for the process are essentially the same as discussed earlier for fresh wastes except that an additional 50% surface area must be added. When using fine screening or septic tanks for pretreatment, it is suggested that no BOD reduction be credited to the pretreatment. Because the amount of BOD removal by pretreatment is relatively small, this procedure provides a slightly conservative basis for design. Selection of an appropriate septic tank retention time, when followed by secondary treatment, is done primarily on the basis of sludge digestion and storage capacity. Field testing has shown that a 12-hr retention time will yield about 1 year of continuous operation before removal of sludge for ultimate disposal. On this basis, a 12-hr septic tank retention time is recommended for most applications.

The size of the flow equalization tank following the septic tank is determined on the basis of the daily wastewater flow pattern. The purpose of the flow equalization tank is to provide a relatively uniform wastewater flow to the process even though almost all of the daily flow may enter the plant over a relatively short time period. The following are general guidelines for determining flow equalization tank capacity:

Daily period when waste-water flow is less than 25% of average flow	Recommended flow equalization tank capacity as percentage of daily flow
Hr	Percentage
0	0
4	10
6	15
8	25
12	33
14	50
16	60
18 or more	67

For applications where soil conditions do not permit excavation for installation of treatment equipment, a package plant installation similar to that shown in Figure 4-25 can be operated with an above-ground septic tank. Operation of the system would be similar to that for the side-by-side configuration with the exception that wastewater would have to be pumped into the septic tank. It is also possible to operate a media assembly in an aboveground position simply by installing the concrete tankage above ground level.

Transportable Plants

Rotating contactor package units are transportable secondary treatment plants which can be used to meet temporary treatment requirements. When operated in conjunction with primary treatment and sludge disposal facilities, they can provide complete secondary treatment at one site for a given period of time and then can be easily moved to another site. Applications for transportable plants include subdivisions awaiting connection to municipal sewer systems and small communities eventually to be connected to regional treatment plants.

Individual Home Treatments Plants

The nature of the construction and operation of the rotating contactor process enables it to be readily scaled up and down over a wide range of equipment sizes with no loss of process efficiency or change in effect of process variables. This fact has been verified in testing over a range of media sizes of 1 to 12 ft in diameter.

The ability to use a wide range of media diameters and shaft lengths allows the process to be applied to a wide range of wastewater treatment applications including the treatment requirements of an individual residence. Use of the rotating contactor process is especially attractive for this application because of its low power consumption and low maintenance requirements. A small treatment plant must be capable of successful operation on a virtually maintenance-free basis, because it will receive little or no operator attention. Operation of the rotating contactor process in combination with a septic tank goes a long way toward achieving this objective. The low power consumption by the process also greatly reduces the incentive of the owner to interrupt power supplied to his treatment system in an effort to reduce operating costs. Field testing of prototype, residential equipment has been conducted[10,11] and design criteria are identical to those of the larger package plants. Effluent produced by the process is suitable for either surface or subsurface disposal.

REFERENCES

1. **Marki, E.,** Results of experiments by EWAG with the rotating biological filter, *Tech. Hochsch. Zurich — Fortbildungskurs der EWAG Hydrol.,* 26, 408, 1964.
2. **Welch, F. M.,** Preliminary Results of a New Approach in the Aerobic Biological Treatment of Highly Concentrated Wastes, Proc. 23rd Purdue Ind. Waste Conf., W. Lafayette, Ind., May 7–9, 1968, 428, W. Lafayette, Ind.
3. **Antonie, R. L. and Welch, F. M.,** Preliminary Results of a Novel Biological Process for Treating Dairy Wastes, Proc. 24th Purdue Ind. Waste Conf., W. Lafayette, Ind., May 6–8, 1969, 115, W. Lafayette, Ind.
4. **Antonie, R. L.,** Nitrification and Denitrification with the BIO-SURF Process, paper presented at the N. Engl. Water Pollut. Control Assoc., Kennebunk, Me., June 10–12, 1974, Kennebunk, Me.
5. **Haug, R. and McCarty, P.,** Nitrification with submerged filters, *J. Water Pollut. Control Fed.,* 44, 2086, 1972.
6. **Bretscher, U.,** Phosphate elimination with RBC's, *GWF Das Gas-Und Wasserfach,* 110(20), 538, 1969.
7. **Weber, E.,** Phosphate Removal with Alum in a Rotating Biological Disc System, project paper for Department of Civil Engineering, Duke University, Durham, N.C., April 1973.
8. **Torpey, W. N. et al.,** Rotating disks with biological growths prepare wastewater for disposal or reuse, *J. Water Pollut. Control Fed.,* 43, 2181, 1971.
9. **Lue-Hing et al.,** Nitrification of a High Ammonia Content Sludge Supernatant by Use of Rotating Discs, paper presented at the 29th Purdue Ind. Waste Conf., W. Lafayette, Ind., May 7–9, 1974, W. Lafayette, Ind.
10. **Otis, R., Hutzler, N., and Boyle, W.,** On-site Household Wastewater Treatment Alternatives — Laboratory and Field Studies, paper presented at the Rural Environ. Eng. Conf., Warren Ver., Sept. 26–28, 1973, Warren Ver.
11. **Homel, J.,** BIO-SURF Septic Tank Treatment Project, interim report to State of Wisconsin Department of Natural Resources; paper presented at the fall seminar of the University of Wisconsin, Green Bay, Nov. 8, 1973.

Chapter 5

UPGRADING EXISTING TREATMENT PLANTS

INTRODUCTION

Many regulatory agencies are beginning to impose more stringent treatment requirements on municipal wastewater treatment plants. Primary treatment plants must upgrade to secondary treatment, and many secondary treatment plants must upgrade to higher flow capacities and higher levels of treatment. In many areas of the U.S., the new requirements also include removal of ammonia nitrogen as well as additional removal of BOD and suspended solids.

The rotating contactor process characteristics of modular construction, low hydraulic head loss, and shallow excavation allow it to be easily integrated with existing treatment facilities in a variety of configurations to upgrade the existing level of treatment.

While the following discussion pertains principally to upgrading existing municipal treatment plants, the basic approach will be applicable to the upgrading of an existing industrial waste treatment plant as well.

UPGRADING EXISTING PRIMARY TREATMENT PLANTS

Conventional Upgrading

Primary treatment plants consisting of primary clarifiers, Imhoff tanks, or septic tanks can be upgraded to secondary treatment standards. This is accomplished by adding shafts of media and a secondary clarifier to follow the existing facility. In some instances, it may be necessary to increase sludge disposal facilities to handle the additional sludge generated by secondary treatment. It is also possible to convert existing primary clarifiers to secondary clarifiers with little or no modification, because the existing solids collection and removal equipment will be suitable for use with a rotating contactor process. Substitution of fine screening for primary treatment will, in many cases, result in significant cost savings and requires a very short construction time.

Subjacent Clarification

Primary treatment plants which have one or more large rectangular primary clarifiers of at least 10-ft sidewater depth can be upgraded by installing shafts of media along the top of the tank as shown in Figure 5-1. A false bottom is then installed just beneath the rotating media to isolate the bottom portion of the tank as a sedimentation zone. Raw wastewater receives pretreatment either in an adjacent settling tank or by fine screening. The pretreated wastewater then enters what was previously the effluent end of the tank and passes through the stages of rotating contactor treatment. The treated wastewater and excess biological solids pass beneath the false bottom for secondary clarification and the clarified effluent is discharged to the existing outfall.

The shallow depth of the subjacent clarifier (4 to 6 ft) is sufficient because of the low solids concentration (100 to 200 mg/l) entering the clarification zone. Surface overflow rate is the primary factor determining solids separation efficiency, and the effective area available is the entire area covered by the false bottom. The small amount of settled solids on the bottom of the tank

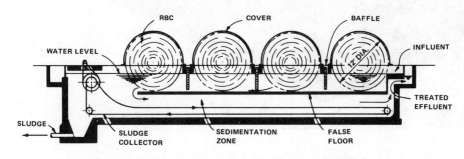

FIGURE 5-1.   Installation of rotating biological contactor and subjacent clarification.

are also not disturbed by the collector mechanism. When existing tankage is not sufficient to meet treatment requirements, additional tankage with or without subjacent clarification can be constructed in parallel with the existing tankage.

The important features of the subjacent clarification design are

· Combines secondary treatment and clarification in existing tanks.
· Reduces construction cost.
· Reduces land area required.
· Reduces hydraulic head loss.
· Permits enclosing both secondary treatment and secondary clarification.

Figure 5-2 shows a full-scale application of this concept (before being covered) at Edgewater, New Jersey, which has been operating successfully since 1973.[1]

The installation of rotating contactor equipment in existing rectangular primary clarifiers requires some modifications to the tanks. In most such tanks, the modifications will consist of the following:

· Lower flight return rail so that collector flights return about 3 ft above tank floor.
· Remove concrete overhang on sidewalls if 1 ft or more.

· Remove scum collection equipment.
· Remove effluent collection launders.
· Remove cross-tank beams if shafts of media cannot be installed to avoid them.
· Install false bottom.
· Add new effluent weir if necessary.
· Add cross-tank beams between shafts of media if necessary.
· Add interstage baffles between adjacent shafts of media.
· Rearrange effluent channels for influent flow distribution.

For subjacent clarifier applications, shaft assemblies can be supplied preassembled to support frames which rest directly on the top of the tank walls. This enables the shafts to be lowered into the tank and operate at the normal water level. For treatment plants which have additional upstream hydraulic head available, the operating water level can be raised and the shaft bearings can then be mounted directly to the top of the tank walls. This eliminates the need for support frames and reduces equipment costs.

Application of the rotating contractor process in existing primary tanks will be most attractive where limited land availability makes conventional upgrading impossible or prohibitively expensive. One important benefit of the rotating contactor is that existing lift stations and outfalls can be used

FIGURE 5-2. Full-scale upgrading of primary treatment at Edgewater, New Jersey.

without alteration. Only the clarifiers require modification. In some cases, sludge handling facilities will have to be expanded to process the additional solids created by secondary treatment.

Although this discussion is directed principally to upgrading existing primary plants, the folded-flow technique just described is also applicable to new plant construction. Significant reductions in construction cost are possible with the subjacent clarifiers, particularly where land is limited and where limited hydraulic head is available.

UPGRADING TRICKLING FILTER PLANTS

Figure 5-3 shows two ways in which a trickling filter plant can be upgraded with the rotating contactor process. When a trickling filter is hydraulically overloaded, rotating contactor equipment can be installed in parallel with the filter and utilize the existing final clarifier. In many instances, trickling filters must be upgraded to levels of treatment which they cannot consistently achieve, especially during cold weather. In this case, it is more economical to install rotating contactor equipment in series with the filter and still utilize the existing secondary clarifier.

In most cases, it is possible to install the new process equipment in series without the need for further pumping, because of the low hydraulic head loss through the rotating contactor process. Shallow excavation will also enable the equipment to fit into the topography without extensive site work. The general nature and concentration of sludge solids in the existing trickling filter effluent will be similar to that leaving a rotating contactor system providing roughing treatment and will, therefore, be compatible with the process.

When upgrading a trickling filter, the rotating contactor process provides the following benefits:

· Provides nitrification and additional BOD removal.
· Treats trickling filter effluent directly without intermediate clarification.
· Utilizes existing secondary clarifier.
· Minimizes site modification.
· Minimizes hydraulic head loss.

In some cases, site conditions will make it necessary to operate the new process equipment following rather than preceding the existing secondary clarifier of a trickling filter plant. In this situation, the suspended solids content of the rotating contactor process effluent should be compared to effluent standards to determine if clarification is required prior to discharge. This will be especially important when upgrading to nitrification standards, because nitrifying bacteria generate relatively little excess sludge.

Use of the rotating contactor process in this application permits full utilization of the existing trickling filter. The trickling filter provides removal of a portion of the carbonaceous matter from the wastewater, which allows nitrifying organisms to predominate on much of the rotating contactor media. This combination permits efficient and economical upgrading to be achieved.

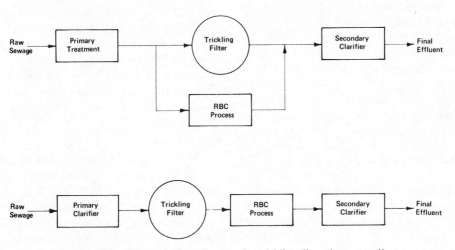

FIGURE 5-3. Schematic flow diagrams for trickling filter plant upgrading.

BOD Removal

Design procedures for upgrading trickling filters are somewhat different from those discussed previously for other applications. For overall design purposes, the rotating contactor process and existing trickling filter are considered as a single process yielding a certain overall degree of treatment.

The required rotating contactor surface area to accomplish the entire secondary treatment task (assuming no trickling filter is presented) is determined as shown in Example #1, Chapter 4. Then, the contactor surface area required to accomplish the treatment being achieved by the existing trickling filter is also determined as in Example #1. This second calculation determines the equivalent contactor surface area which the trickling filter represents. Subtracting the second calculated surface area from the first yields the amount of new process surface area which must be added to achieve the required overall degree of treatment. A typical design is calculated in Example #5.

Example #5 -- For a 1.0-mgd. wastewater flow, the BOD concentration is 200 mg/l. Primary treatment achieves 25% BOD removal and an effluent of 150 mg/l. The existing trickling filter achieves 75% BOD removal and an effluent of 38 mg/l; a 10 mg/l BOD final plant effluent or 95% overall plant BOD removal is required. BOD removal by secondary treatment must be 93.3%. From Figure 4-1 the required hydraulic loading on the rotating contactor at 150 mg/l BOD and 93.3% BOD removal is 1.7 gpd/ft^2. The required surface area is then $\frac{1,000,000 \text{ gpd}}{1.7 \text{gpd/ft}^2}$ = 588,000 ft^2 (assuming no trickling filter is present). To obtain the 75% BOD removal achieved by the existing trickling filter at 150 mg/l BOD, the process would have to be loaded at about 7 gpd/ft^2. For this application, the trickling filter is equivalent to:

$\frac{1,000,000 \text{ GPD}}{7 \text{ GPD/ft}^2}$ = 143,000 ft^2 of surface area. The net rotating contactor process surface area required for upgrading is then

588,000 – 143,000 = 445,000 ft^2

When conditions require a design for low wastewater temperature, the procedure shown in Example #2, using Figure 4-3, is followed to make the proper adjustment.

Example #6 – For the same application as Example #5, a design for wastewater temperature

of 45° F is necessary. For 93.3% secondary treatment and 45°F, the temperature correction factor from Figure 4-3 is 1.95. For the 75% treatment level obtained by the trickling filter, there is no correction factor shown in Figure 4-3. Using the correction factor for the lowest level of treatment shown will be a good estimate. In this case, for 82% treatment and 45°F, the correction factor is 2.2. This correction factor is used to reduce the equivalent amount of surface of the trickling filter, because it will exhibit reduced treatment capacity at low temperatures similar to the rotating contactor process. Net surface required is then

$588,000 \times 1.95 - \frac{143,000}{2.2} = 1,147,000 - 65,000 =$ 1,082,000 ft^2.

If the BOD removal by the trickling filter at the low wastewater temperature was known to be 75%, then the temperature correction factor would be used to increase the equivalent amount of surface area.

Example #7 – For the same application as Example #6, the trickling filter obtained 75% BOD removal at 45°F. The net surface area required is then

$588,000 \times 1.95 - 143,000 \times 2.2 = 1,147,000 - 315,000$ = 832,000 ft^2.

For applications where the primary effluent BOD is approximately 150 mg/l and the wastewater temperature is 55°F or above, Figure 5-4 can be used directly to determine required process surface area. The ordinate indicates BOD removal by combined secondary treatment of the trickling filter and rotating contactor process. Figure 5-4 was developed using the design procedure described by Example #5. The hydraulic loading line for 0% BOD removal by the trickling filter is the same as for 150 mg/l BOD wastewater seen in Figure 4-1.

If the wastewater temperature is below 55°F, or if the primary effluent BOD is significantly different from 150 mg/l, it will be necessary to calculate the required amount of process surface area by the longer method described in Examples #5, 6, and 7. A correction factor cannot be applied to the rotating contactor surface area requirement calculated with Figure 5-4, since the

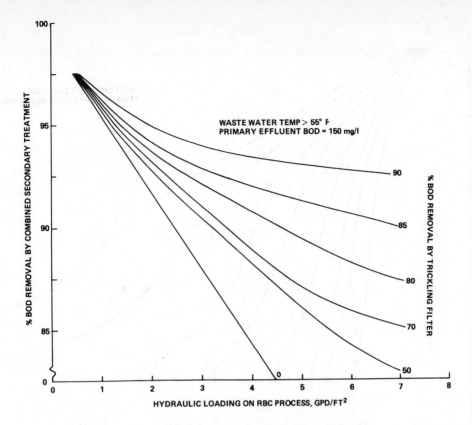

FIGURE 5-4. Design correlation for upgrading trickling filters.

equivalent surface area of the trickling filter must also be adjusted.

Nitrogen Control

When upgrading a trickling filter for nitrification as well as BOD removal, the two-step calculation procedure should also be employed using Figure 4-7 and the procedures outlined in Example #5. The larger of the two surface area determinations should then be used for design. For applications of 150 mg/l BOD and 55°F primary effluent, Figure 5-5 can be used to determine the required surface area directly.

Trickling filter plants can also be upgraded to provide denitrification. This would be done as shown in Figure 5-6. Trickling filter effluent would go directly to rotating contactor equipment for additional BOD removal and nitrification as described above. Nitrified effluent would then flow to completely submerged media for denitrification as described in Chapter 4. Denitrified effluent would then flow to the existing final clarifier, which would be the only clarifier in the process. Very little head loss would occur through

the nitrification and denitrification steps, so that additional wastewater pumping will normally not be required.

UPGRADING ACTIVATED SLUDGE PLANTS

Many states have established effluent standards based on ammonia nitrogen and total oxygen demand as well as BOD and suspended solids. This means that many activated sludge plants must be upgraded to meet these new standards, because previous design practices considered only BOD and suspended solids removal.

The rotating contactor process can be applied to upgrading an activated sludge plant to achieve nitrification in several ways as shown in Figure 5-7. An activated sludge plant presently providing only BOD reduction can be upgraded also to achieve nitrification by reducing the influent flow rate. This extends the wastewater retention time and allows a sludge age to develop which will achieve nitrification within the single stage of aeration. The balance of the flow is then treated

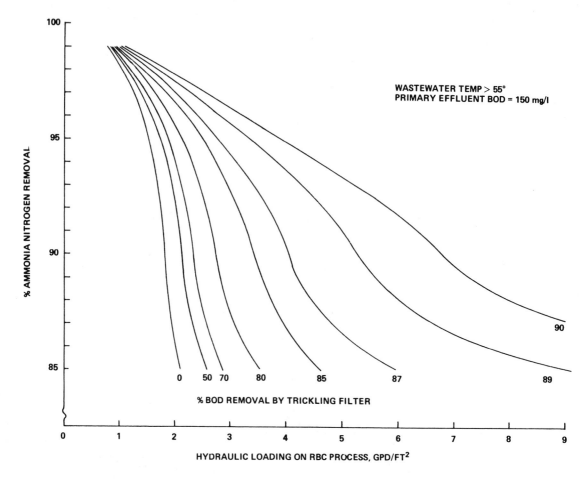

FIGURE 5-5. Design correlation for upgrading trickling filters for nitrification.

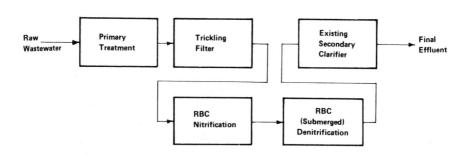

FIGURE 5-6. Schematic flow diagram for trickling filter upgrading including nitrification-denitrification.

by rotating contactor equipment installed to operate in parallel with the existing aeration tanks and designed to provide both BOD removal and nitrification. Rotating contactor effluent is then directed to the existing secondary clarifier. If the secondary clarifier has sufficient capacity for the total hydraulic flow, it need not be expanded.

In another mode of operation, the rotating contactor process can be used to remove carbon-aceous BOD preceding an existing activated sludge plant. The partially treated wastewater along with the excess solids are discharged directly into the existing activated sludge system. The aeration, clarification, and sludge recirculation systems are then operated to achieve nitrification. This type of operation has undergone pilot plant studies[2] and has been found to work quite well. The addition of the solids produced by the rotating contactor

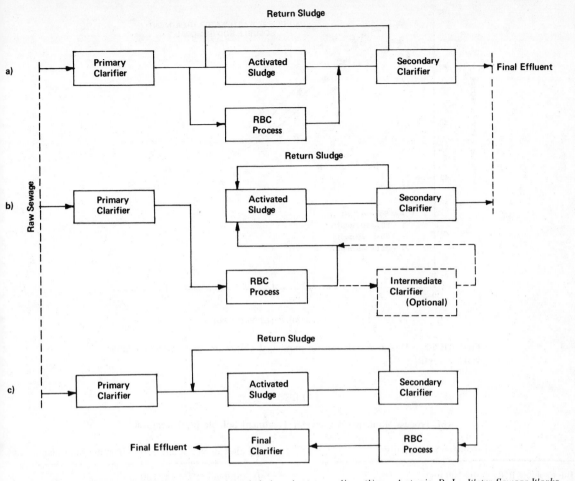

FIGURE 5-7. Schematic flow diagrams for activated sludge plant upgrading. (From Antonie, R. L., *Water Sewage Works*, 121(11), 44, 1974; 121(12), 54, 1974. With permission.)

process to the activated sludge system improved the settleability of the suspended nitrifying culture so that stable nitrification could be achieved.

Figure 5-8 and Table 5-1 can be used to design for partial soluble BOD removal prior to an existing aeration system. Soluble BOD removals of 40 to 50% have been found to provide sufficient pretreatment for stable nitrification and good solids settleability.[2] Figure 5-8 indicates loadings in the range of 6 to 10 gpd/ft^2 for this application.

The third way to upgrade an activated sludge plant to achieve nitrification is to install the rotating contactor equipment after the existing secondary clarifier. When operated after the secondary clarifier, a nitrifying culture will predominate over the entire surface area of the media. The effluent after nitrification will contain a relatively low level of sludge solids. If the existing activated sludge plant is producing effluent BOD, suspended solids, and ammonia nitrogen concentrations, each measuring 20 mg/l, the rotating

contactor process effluent after complete nitrification will show no significant increase in suspended solids concentration. This results from a very low rate of growth by the nitrifying culture and from the offsetting factors of endogenous respiration and the effect of higher life forms, such as protozoans and rotifers, acting as predators. In many such applications, the effluent can go directly to the receiving body, if the level of suspended solids produced by the activated sludge plant is still suitable for discharge. If not, then the effluent can pass directly to tertiary filtration.

Direct filtration of the effluent has been tested in pilot plant scale[2] and the results are shown in Table 5-2. The tests show that effluents of 2 mg/l suspended solids can be consistently produced at economical filter loading rates and cycle times. When a more moderate effluent solids level is required, e.g., 10 mg/l, then a reactor-clarifier with chemical addition could be substituted for the filtration system.

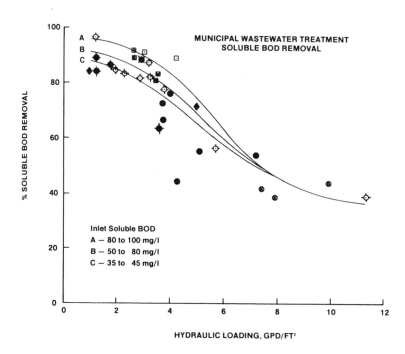

FIGURE 5-8. Design correlation for municipal wastewater pretreatment soluble BOD removal.

TABLE 5-1

RBC Process Municipal Wastewater Treatment Soluble BOD Removal

0.5-m-diameter pilot plants		Full-scale plants and 3.2-m-diameter pilot plants	
Inlet soluble BOD = 80–100 mg/l		Inlet soluble BOD = 80–100 mg/l	
□–Philadelphia, Pa.	83–88 mg/l	□–Philadelphia, Pa.	86–96 mg/l
Inlet soluble BOD = 50–80 mg/l		Inlet soluble BOD = 50–80 mg/l	
□–Philadelphia, Pa.	77 mg/l	□–Philadelphia, Pa.	72–77 mg/l
		O–Gladstone, Mich.	53 mg/l
		O–Indianapolis, Ind.	64–74 mg/l
Inlet soluble BOD = 35–45 mg/l		Inlet soluble BOD - 35–45 mg/l	
◆–Shelbyville, Ind.	35–40 mg/l	●–Gladstone, Mich.	44 mg/l
		●–Indianapolis, Ind.	35–45 mg/l
		■–Niagara, Wis.	36 mg/l

When comparing the addition of rotating contactors to an existing activated sludge plant against the addition of a second step of aeration, clarification and sludge recycle, the rotating contactor approach will eliminate the need for the second clarification step. This holds true whether the effluent can be directly discharged or must be filtered.

Pilot Plant Tests

Rotating contactor pilot plants of various sizes have been tested at a number of municipalities around the U.S. on activated sludge effluents.[3] Units containing 0.5-m-diameter media with a nominal capacity of 500 gpd were tested at Mansfield, Ohio, Tiffin, Ohio, and Broward County, Florida. A 2-m-diameter pilot plant with a nominal capacity of 20,000 gpd was tested at Madison, Wisconsin. Figure 5-9 is a 3.2-m-diameter unit with a nominal capacity of 50,000 gpd. This unit was tested at Phoenix, Arizona. All the pilot plants have four stages of media assembled onto a single shaft. Bulkheads in the tankage containing the media separate the four sections into

TABLE 5-2

Gravity Filtration of Rotating Biological Contactor Effluent

Dates 1974	No. of runs	Filter Media	Flow gpm/ft²	Average solids loading lb/ft²	Average solids load to H = 4 ft lb/ft²	Length of runs hr	Head loss ft	TSS Influent mg/l	TSS Effluent mg/l	Total BOD (inhibited*) Influent mg/l	Total BOD (inhibited*) Effluent mg/l	Effluent turbidity JTU
4/17–5/3	19	24 in. Coal 12 in. Sand 9 in. Gravel	2.46	0.62	0.53	9.30 hr	4.64 ft	21	1.9	18	4	2.0
5/3–5/7	6	14 in. Coal 9 in. Sand 9 in. Gravel	2.52	0.55	0.35	14.64 hr	6.17 ft	18	2.0	20	8	2.7
5/7 –5/10	7	Same as above	3.83	0.69	0.46	10.70 hr	6.04 ft	15	2.0	18	5	2.3
5/10–5/14	6	6 in. Low density Coal 14 in. Coal 9 in. Gravel 9 in. Sand	4.00	0.84	0.50	14.22 hr	6.96 ft	18	1.7	15	2	1./
5/14–5/22	4	Same as above	2.47	0.86	0.47	40.96 hr	7.30 ft	29	2.8	21	5	2.0
5/23–5/28	11	Same as above	4.72	1.42	0.93	12.19 hr	6.11 ft	25	1.5	18	4	1.7
6/4–6/6	3	Same as above	3.77	0.72	0.47	20.58 hr	6.14 ft	16	1.4	19	5	2.0
6/6 –6/10	3	Same as above	2.52	0.55	0.33	22.41 hr	6.70 ft	18	1.4	21	5	2.0
6/10–6/12	7	Same as above	5.89	1.82	1.48	8.74 hr	4.90 ft	26	2.6	25	6	2.6

*Inhibitor add to suppress nitrification during incubation.

TSS – Total suspended solids.

FIGURE 5-9. 50,000-gal/day pilot plant unit. (From Antonie, R. L., *Water Sewage Works,* 121(11), 44, 1974; 121(12), 54, 1974. With permission.)

individual stages of treatment. Activated sludge effluent is pumped at a relatively constant rate by the rotating bucket pump mechanism into the first stage of media, and it flows by gravity through the four stages in series.

Figure 5-10 shows test data from these five pilot plant programs. The data shown are from samples of individual stages of treatment from all the pilot units. Most of the data are from grab samples, some of which were collected at appropriate intervals to allow wastewater retention time through the subsequent stages of treatment. Some data are from composite samples taken over periods varying from 4 to 24 hr. Variation in sampling procedures is responsible for part of the scatter in the data. Since hourly fluctuations in ammonia content can be quite significant, the grab sampling techniques would show significant variation from stage to stage and from day to day. Part of the scatter is attributable to variations in the treatability of the waste at the various locations due to differences in alkalinity, contributions of inhibitory substances from industry, and changes in activated sludge plant operation. This points out the importance of conducting pilot plant studies for nitrification applications, because when these variable factors are present, they can significantly affect full-scale plant design. The line drawn through the data points was drawn through the lower range of values to arrive at a reasonably conservative basis for design.

Design Procedure

The activated sludge effluents treated during these tests generally had BOD concentrations of 20 mg/l or less and soluble carbonaceous BOD concentrations of 10 mg/l or less. Because of this, nitrifying bacteria were predominant on the rotating media, and oxidation of ammonia nitrogen was the only significant reaction occurring. This allows the development of a generalized design procedure which can cover a wide range of operating conditions. The design procedure which follows is commonly used in design of unit operations such as distillation and extraction where a physical equilibrium based on solubility or vapor pressure exists between components. The procedure also lends itself well to other staged operations where a similar kind of equilibrium can be defined. This method utilizes the correlation in Figure 5-10 for nitrification of secondary effluent with the rotating biological contactor process. The line drawn represents a biochemical equilibrium between the rate at which the fixed biological culture is oxidizing ammonia nitrogen, and the concentration of ammonia nitrogen it is being exposed to.

Figure 5-11 contains a reproduction of the equilibrium relationship and was used to develop what is called an operating line. If we arbitrarily select a surface loading rate on an individual stage of treatment (i.e., 8 gpd/ft^2) we can determine how much ammonia nitrogen will be removed at

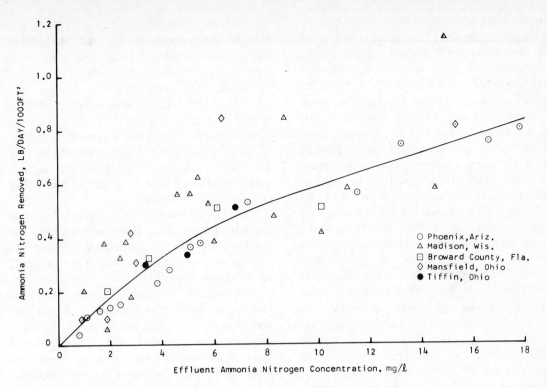

FIGURE 5-10. Operating data correlation for single stage ammonia nitrogen removal. (From Antonie, R. L., *Water Sewage Works,* 121(11), 44, 1974; 121(12), 54, 1974. With permission.)

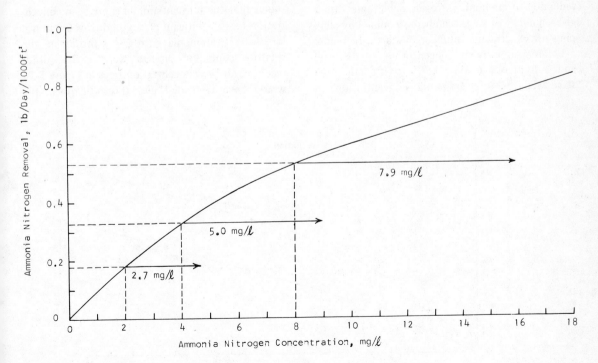

FIGURE 5-11. Construction of operating line for 8.0 gpd/ft² hydraulic loading. (From Antonie, R. L., *Water Sewage Works,* 121(11), 44, 1974; 121(12), 54, 1974. With permission.)

various effluent ammonia nitrogen concentrations. For example, if the fixed biological film is exposed to 8 mg/l of ammonia nitrogen, it will remove approximately 0.54 lb of ammonia nitrogen/day/1,000 ft^2 of media surface area. At 8 gpd/ft^2, this results in a reduction of 7.9 mg/l. This means the influent ammonia nitrogen strength to that stage was approximately 16 mg/l. Similarly, when exposed to an ammonia nitrogen concentration of 4 mg/l, the fixed nitrifying culture will remove 0.33 lb/day/1,000 ft^2. At 8 gpd/ft^2, 5 mg/l will be removed. At 2 mg/l concentration, 0.19 lb/day/1,000 ft^2 will be removed for a reduction of 2.7 mg/l. If a line is then drawn from the intersection of the coordinates through the three arrowheads in Figure 5-11, we have an operating line. This line is shown along with the equilibrium line in Figure 5-12. The equilibrium line represents

The equilibrium line represents the locus of effluent concentrations from the stages, and the operating line represents the locus of influent concentrations to the stages. Thus, the number of stages required to achieve a given reduction of ammonia nitrogen can be determined by stepping off of stages with a series of horizontal and vertical lines, as in Figure 5-12. Alternatively, the reduction of ammonia nitrogen, achievable with a given number of stages can be determined by stepping off exactly that number of stages. In Figure 5-12, for an influent concentration of 18 mg/l and a hydraulic loading of 8 gpd/ft^2, the first stage reduces the ammonia concentration to

approximately 10 mg/l. In the second stage, it is reduced to approximately 4.5 mg/l, in the third stage to approximately 2.1 mg/l, and in the fourth stage to approximately 1 mg/l. With four stages of operation, the overall hydraulic loading is 2 gpd/ft^2, and the total ammonia nitrogen reduction is from 18 mg/l to 1 mg/l. The operating line can now be used to design a multistage system, each of which is operated at the same hydraulic loading. The system can be designed for any influent ammonia strength and to produce any desired effluent ammonia strength.

Figure 5-13 shows operating lines for other hydraulic loading rates. These operating lines were developed with the same procedures as Figure 5-11 and can be used in a similar manner as Figure 5-12 to determine ammonia nitrogen reduction for various influent ammonia concentrations and numbers of stages. The procedure of Figure 5-12 was repeated many times for various ammonia nitrogen concentrations and various overall hydraulic loading rates for four stages of operation using Figure 5-13. The results are shown in Figure 5-14, where percent ammonia nitrogen removal is shown as a function of overall hydraulic loading and for various influent ammonia strengths. It shows that the influent ammonia nitrogen concentration has a significant effect on nitrogen removal kinetics. If ammonia nitrogen removal in the rotating contactor process were a first order reaction, all the concentration lines in Figure 5-14 would have coincided. For this to happen, the

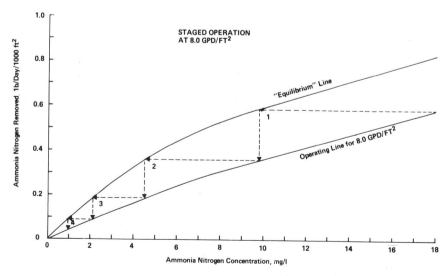

FIGURE 5-12. Illustration of design procedure for staged operation. (From Antonie, R. L., *Water Sewage Works*, 121(11), 44, 1974; 121(12), 54, 1974. With permission.)

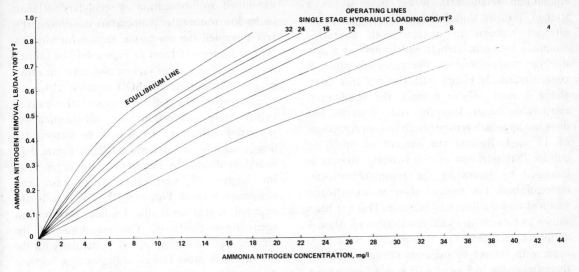

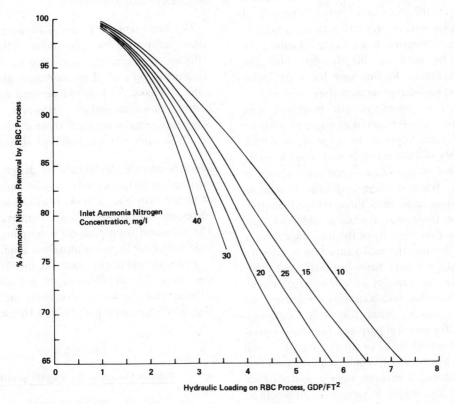

FIGURE 5-13.   Nitrification design procedure diagram. (From Antonie, R. L., *Water Sewage Works*, 121(11), 44, 1974; 121(12), 54, 1974. With permission.)

FIGURE 5-14.   Design correlation for nitrification of secondary effluent: four stages. (From Antonie, R. L., *Water Sewage Works*, 121(11), 44, 1974; 121(12), 54, 1974. With permission.)

equilibrium relationship would have to be a straight, positive sloped line, indicating ammonia nitrogen removal to increase with increasing ammonia nitrogen strength indefinitely at a rate directly proportional to the ammonia nitrogen concentration. In Figure 5-10, this was true up to about 6 mg/l. Above 6 mg/l, the equilibrium relationship departs from first order behavior, but does not approach zero order up to concentrations of 18 mg/l. Because the amount of nitrifying culture that develops on the rotating surfaces is increased by increasing the ammonia nitrogen concentration, the amount of ammonia nitrogen removed also continues to increase. This has been shown to be true up to a concentration of 70 mg/l ammonia nitrogen.[4] If the process were zero order with respect to ammonia nitrogen removal, the relationship in Figure 5-10 would have been a horizontal line. While ammonia nitrogen removal may be close to zero order when measured on a per-unit biomass basis $\frac{mg/l\ NH_3-N}{mg/l\ VSS}$, the system does not exhibit zero order kinetics because it develops a biomass in proportion to the ammonia nitrogen concentration being treated. Figure 5-14 can now be used to directly determine the required hydraulic loading rate for a particular nitrification application on secondary effluent.

The design procedures just described were repeated for various numbers of stages in series for degrees of nitrification in the range of 90 to 95%. The capacity of various numbers of stages in series, as compared to four-stage operation is shown in Table 5-3. When dividing a specific amount of media surface area into three rather than four stages, the three-stage system would have only 90% of the flow capacity of the four-stage system. Similarly, dividing the media into two or into just a single stage would have just 80% and 60%, respectively, the capacity of a four-stage system. Dividing the media into six stages would have just 7% more capacity than four-stage operation. Therefore, for practical purposes, four-stage operation is recommended. For applications to small treatment plants, which would require just a few shafts of media, it may be more practical to use fewer than four stages. In those cases, it would be necessary to increase the amount of surface area determined from Figure 5-14 by the factor shown in Table 5-3.

Wastewater temperatures below 55°F affect nitrification efficiency. However, relatively little test data are available in this temperature range for secondary effluents. Until additional test data are

developed on nitrification of secondary effluents under low wastewater temperature conditions, it is recommended the correction factors for nitrification of primary effluent in Figure 4-10 be utilized for design. Since these factors also take the effect of low temperature on BOD removal efficiency into account, they will be a conservative basis of temperature correction for nitrification of secondary effluents. To use these factors, the design loading rate determined from Figure 5-14 would be divided by the appropriate factor based on degree of nitrification and wastewater temperature from Figure 4-10. This yields the required design hydraulic loading for the low temperature condition. The magnitude of the temperature correction factors, particularly for the lowest temperature ranges, indicates how costly it is to achieve nitrification for low wastewater temperatures. Under such conditions, the necessity for maintaining the same degree of nitrification must be carefully considered.

The test data in Figure 5-10 were collected under uniform flow conditions. Therefore, if effluent requirements are based on hourly sampling or on peak flow conditions, the relationships in Figure 5-14 should be used for peak or hourly flow conditions. However, if daily composite samples are used, there will be little loss in nitrification efficiency due to flow variations; and average flow can be used for design. Although nitrification efficiency will decrease during periods of higher than average flow, it will increase during corresponding periods of lower than average flow. The net result from composite sampling will show little difference from uniform flow conditions.

Full-scale municipal plants in the 2- to 3-mgd size range for nitrification of activated sludge effluent are now in operation at Cadillac, Michigan,[5] Sarasota, Florida, and Hinckley, Ohio.

TABLE 5-3

Relative Capacities for Staged Operation

(90–95% Nitrification)

No. of stages	Relative capacity
1	0.60
2	0.80
3	0.90
4	1.00
6	1.07

Denitrification

Activated sludge plants can be upgraded to provide both nitrification and denitrification as shown in Figure 5-15. Clarified secondary effluent would be treated by partially submerged rotating contactors for nitrification and then flow directly to submerged media for denitrification. Because denitrification with methanol produces relatively little excess sludge (about 10 mg/l), it may still be possible to go directly to tertiary filtration with the denitrified wastewater. This would eliminate two steps of clarification when compared to the alternative of two suspended growth systems for nitrification and denitrification. The design procedure for denitrification would be the same as in Chapter 4, assuming complete conversion of ammonia to nitrate.

UPGRADING EXISTING LAGOONS

Lagoons which are not meeting treatment requirements due to overloading or a change in treatment requirements can be upgraded by the addition of rotating contactor process equipment. Shafts of media and a clarifier can be installed in series with such lagoons as shown in Figure 5-16a. This arrangement is suitable for upgrading a lagoon to a higher degree of treatment. Higher degrees of treatment can consist of lower effluent soluble BOD and nitrification.

If the effluent BOD is accounted for principally by algae and other suspended solids, it will pass through the rotating contactor process unaffected. Soluble BOD and ammonia nitrogen are the only components that will be removed. In cases, where only nitrification is required, the effluent BOD and suspended solids concentration from the process may be acceptable for discharge and the clarifier in Figure 5-16a can be omitted. Design procedures for upgrading lagoons for additional BOD removal are identical to those for upgrading trickling filter plants. Design procedures for nitrifi-

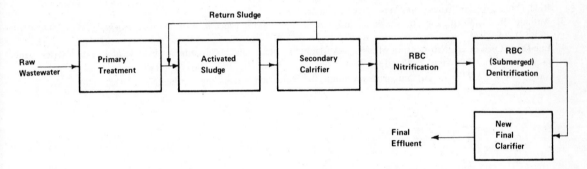

FIGURE 5-15. Schematic flow diagram for activated sludge upgrading including nitrification-denitrification.

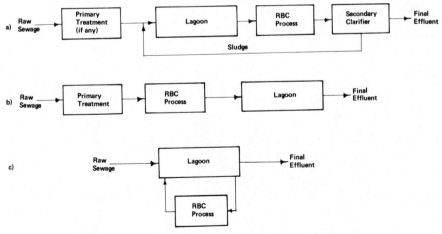

FIGURE 5-16. Schematic flow diagrams for upgrading lagoons.

cation are identical to those for upgrading activated sludge plants.

During winter operation, lagoon effluents are often extremely cold and biological treatment is not feasible. In these cases it is better for rotating contactor equipment to precede the lagoon where the wastewater is much warmer.

An organically overloaded lagoon can be upgraded as shown in Figure 5-16b. In this case, rotating contactor process equipment precedes the lagoon. Primary treatment of the wastewater is required before rotating contactor treatment. Pre-treatment should consist of a conventional primary clarifier or at least grit removal (if necessary) and fine screening. Rotating contactor process solids can pass directly into the lagoon.

Figure 5-16c is an alternate arrangement for lagoon effluent upgrading. In this case, the lagoon contents are recycled through the rotating contactor unit. This arrangement is most suitable for nitrification of the lagoon contents and permits development of higher life forms, especially fish, in the lagoon. Design procedures are identical to those for nitrifying an activated sludge effluent.

REFERENCES

1. **Dobrowolski, F., Brown, J., and Bradley, F.,** Testing Rotating Biological Contactors for Secondary Treatment in a Converted Primary Tank, paper presented at the Annu. Meeting of the Water Pollut. Control Assoc. of Pennsylvania, Aug. 8, 1974.

2. Advanced wastewater pilot plant treatment studies conducted for the Consolidated City of Indianapolis, report by Reid, Quebe, Allison, Wilcox, and Assoc., Inc., January 1975.

3. **Antonie, R. L.,** Nitrification of activated sludge effluent: Bio-Surf process, *Water Sewage Works,* 121(11), 44, 1974; 121(12), 54, 1974.

4. **Lue-Hing, Cecil et al.,** Nitrification of a High Ammonia Content Sludge Supernatant by Use of Rotating Discs, paper presented at the 29th Annu. Purdue Ind. Waste Conf., West Lafayette, Ind., May 7–9, 1974.

5. **Singhal, A. K.,** AWT plant cuts nutrients economically, Water and Wastes Engineering, Nev., 1975, p. 89.

Chapter 6

DESIGN CRITERIA – INDUSTRIAL WASTEWATER TREATMENT

PRETREATMENT

Application of the rotating biological contractor process to industrial wastewater treatment is done in the same fashion as other biological treatment methods. There are requirements for maintaining wastewater pH in the range of 6.5 to 8.5, temperature in the range of 55 to 95°F, and a nutrient balance of BOD:N:P at 100:5:1, which are required for all biological treatment. While the process can operate effectively when these conditions are not met, the capacity of its process equipment is somewhat reduced. A biological process can often adapt to conditions beyond the recommended ranges; however, in industrial waste treatment, variation of these conditions prevents adaptation and is the principal reason for controlling wastewater characteristics.

For most applications, the simplest and most economical means of control utilizes an equalization step. This permits variations in pH, temperature, and nutrient balance to be averaged over a period of time to reduce requirements for chemical addition or wastewater cooling. A variable capacity equalization step allows flow equalization, which is often necessary to equalize flow variations due to shift operation for industrial operations. It also provides a source of organic material during weekend and holiday shutdowns to maintain an active biological culture.

The rotating contactor process has a fixed biological growth and therefore is less susceptible to upset from fluctuations in operating conditions than a suspended growth system. This permits minimizing the size of equalization facilities. However, when necessary, flow equalization not only benefits biological treatment, but also allows more economical design of pumping and piping systems as well as clarifiers and any subsequent treatment facilities. When aerated at a rate of about 50 SCF/lb BOD load, the equalization step will be kept fresh. Aerating at a rate much greater than this can sometimes result in the development of a highly dispersed, poorly settling biological floc, which can carry through the entire treatment system and increase effluent BOD and solids levels. If the recommended aeration rate does not adequately mix the equalization tank, mechanical mixing should be added as a supplement.

In addition to improving treatment efficiency, another reason for controlling wastewater pH is to avoid developing undesirable microbial species. At low pH values, below 6.0, the predominance of yeasts and fungi over bacteria is favored. At still lower pH values, less than 5.0, a predominantly yeast and fungal culture can produce objectionable odors, especially when treating high carbohydrate wastes from food-processing operations. The rotating contactor process has been tested on a wide range of industrial wastes, some as concentrated as 15,000 mg/l BOD, without experiencing plugging of the corrugated media. At continuous, very low pH conditions, however, the attached growth can become quite thick – as much as ½ in. While this will not necessarily plug the media, it will exert extreme stresses on the rotating contactor assembly and could result in mechanical problems if continued indefinitely.

Because many industrial wastes contain relatively little settleable matter, primary clarification is usually not necessary, and primary treatment can be eliminated or consist simply of fine screening.

For installations already having primary clarification, the substitution of primary screening allows existing primary clarifiers to be converted for secondary clarifier use. This can be done without significant modifications to accommodate the rotating contactor solids. Dense, low volume secondary sludge is produced by the process and there is no need to recycle secondary sludge. Therefore, the existing mechanical collector system and sludge pumps will usually be of sufficient type and capacity.

For some cases, however, primary clarification can be important. The use of primary clarification will anticipate variation in the amount of settleable matter for applications where there is relatively little information on the solids content of the wastewater. More important, if BOD loads exceed those used for design of the treatment process, they can often be reduced to design level with chemical addition and clarification. A primary clarifier should also be considered as an alternative for increasing treatment capacity in the future for applications where changes in treatment requirements are expected. A primary clarifier also provides some flexibility in solids handling because secondary solids can be recycled to the primary clarifier for additional thickening before dewatering and disposal.

Oil and grease concentrations of 200 mg/l of animal and vegetable origin and 375 mg/l of mineral origin have been absorbed by the process during pilot plant testing with no loss in treatment efficiency. If anticipated oil and grease levels are above these limits, pretreatment should include their removal. This can be done with any of the commercially available air flotation processes.

VARIABLE WASTEWATER FLOW

One of the problems in providing biological treatment to industrial wastes is the pattern of wastewater flow from industrial operations. Shift operation, holidays, and weekends create the problem of operation with little or no wastewater flow. Cyclic operations, spills, and clean-up periods present requirements for treatment plant operation with varying flows and hydraulic and organic surges.

Intermittent Flow

Pilot plant testing of the rotating biological contactor process has been conducted on intermittent flow operation. Figure 6-1 is a process flow sheet of the pilot plant used. A synthetic wastewater was used for the tests, which consisted of dairy whey solids, $(NH_4)_2 HPO_4$ and $K_2 HPO_4$ mixed in tap water in the ratio of 1.0:0.095:0.70 by weight. Dairy whey provided a source of carbohydrate and organic nitrogen, while di-ammonium orthophosphate provided phosphorus and inorganic nitrogen, and di-potassium orthophosphate acted as a buffer. Waste was made up in concentrated form (up to 100,000 mg/l whey solids) every few days in a continuously mixed and water cooled ($10°C$) 500-gal capacity storage tank. It was delivered by a controlled volume pump to a mixing drum and diluted to the desired operating concentration (500 to 1,300 mg/l COD) by tap water at $70°F$. Constant temperature dilution water was obtained by mixing hot and cold tap water with a thermostatically controlled solenoid valve regulating the hot water flow. Once diluted, the waste was pumped through a rotameter to the first rotating contactor unit. The waste flowed over weirs to get to each successive treatment and settling stage and finally into an effluent trough which led to the drain. Solids which settled out in the settling tanks

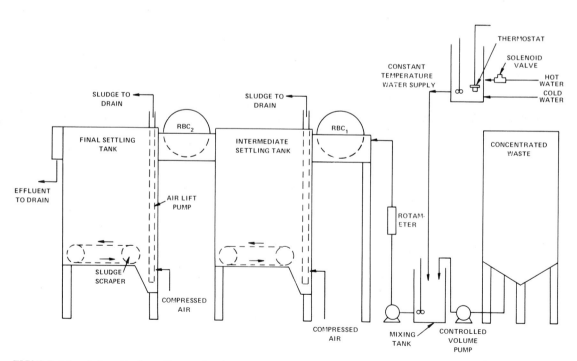

FIGURE 6-1. Schematic flow diagram for simulated industrial wastewater treatment tests. (From Welch, F. M., *Water Wastes Eng. Ind.*, July/August, 1969. With permission.)

were deposited in hoppers by sludge scrapers and sent to the drain by air lift pumps. For the tests described here, the intermediate settling tank was by-passed with a set of troughs to prevent a lag in the response of the second stage of discs.

Each treatment stage contained 87, 3-ft-diameter aluminum discs in 120-gal capacity tanks. The discs were spaced at ½ in. intervals and rotated at controlled speeds. Both units were inoculated with sludge from a local treatment plant. Growth developed on the discs within a week and, at steady state, had a shaggy appearance as shown in Figure 6-2.

Grab samples were taken of the inlet waste-water and of the mixed liquor leaving each stage of discs. Mixed liquor samples were allowed to settle for ½ hr to simulate clarifier operation and the supernatant was withdrawn for analysis. COD analyses were conducted on all samples according to Standard Methods. BOD analyses were conducted occasionally to establish a relationship between BOD and COD for this wastewater, and it

FIGURE 6-2.   Biological growth on RBC unit treating simulated industrial wastewater.

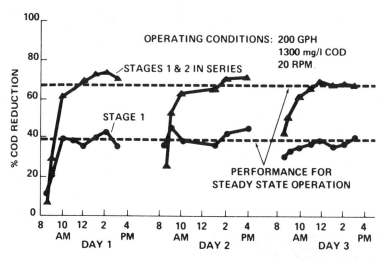

FIGURE 6-3.   Performance with intermittent flow.

was found to be 0.66 mg/l BOD = 1.0 mg/l COD. Field experience indicates that the BOD/COD ratio on dairy waste will vary from 0.7 on raw waste to as low as 0.15 on effluent which has undergone a high degree of treatment.[1]

To simulate industrial operation with intermittent flows, the system was subjected to a wastewater flow only during the regular 8-hr working day. The discs were allowed to rotate overnight in their mixed liquor to prevent the biomass from drying or becoming anaerobic, but there was no wastewater flow. Performance under these conditions is presented in Figure 6-3. Notice the lag in percent COD reduction the first few hours in the morning. Overnight there was gradual sloughing of a small fraction of the solids on the discs into the mixed liquor. This raised the suspended solids concentration to 5 times the normal operating level of 100 to 300 mg/l for a 1,300 mg/l COD wastewater. Continued mixing and aeration overnight appeared to decrease their settleability the next morning and increased effluent BOD values.

When wastewater flow was reinstated the following morning, several hours were required before the excess mixed liquor solids could be washed out and the microbial population brought back to a normal state of activity. Steady state performance under these conditions of flow, concentration, disc speed, and with no intermediate settling are: Stage 1, 38% COD reduction; Stage 1 and Stage 2 in series, 66% COD reduction. To avoid build-up of mixed liquor solids, a low flow of wastewater or tap water and low disc RPM were

maintained overnight. These test results are presented in Figure 6-4. Note that the initial period of operation showed greater COD removal than experienced for steady state operation. This occurred because the low mixed liquor COD levels produced overnight kept the effluent COD concentration low for a few hours the following morning. Under actual treatment plant operation, the effluent from the final clarifier can be recycled to provide removal of sloughed organisms during periods of no wastewater flow. The soluble organic materials remaining in the clarifier are then used to maintain the attached biological culture in a higher state of activity. This also prolongs the period of low effluent concentration the following morning until the contents of the final clarifier are displaced.

The rotating contactor process functions effectively under conditions of intermittent flow and will maintain steady state performance as long as a low wastewater flow or effluent recycle is utilized under plant shutdown conditions.

Hydraulic and Organic Surges

Hydraulic and organic surge tests have been conducted by Bennett et al.[2] on pulp and paper waste. The pilot plant RBC unit used is identical to that of Figures 3-18 and 6-5 and is described in Chapter 3.

A schematic diagram of the pilot plant system is shown in Figure 6-5. Wastewater from an insulating board mill was passed through a 0.020-in. screen ("Hydrasieve") and flowed to a constant level mixing tank. Nitrogen and phosphorus nu-

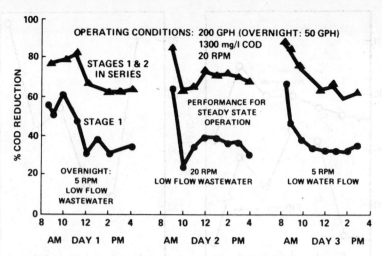

FIGURE 6-4. Performance with intermittent flow and low flow overnight.

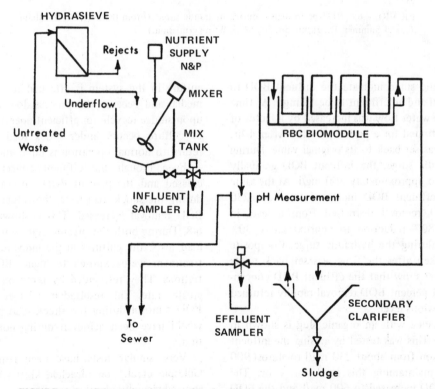

FIGURE 6-5. Schematic diagram of hydraulic and organic surge pilot plant. (From Bennett, D., Needham, T., and Summer, R., *Tappi*, 56, 50, 1973. With permission.)

trients were added to the tank by gravity flow in the form of ammonium hydroxide and phosphoric acid to obtain a BOD:N:P ratio of 100:5:1. Flow from the mixing tank passed through pH measurement to the wet well of the rotating contactor unit. No attempt was made to control pH. A small,

conical, stainless steel clarifier was attached to the end of the unit. Composite samples were taken of the rotating contactor influent and clarifier effluent.

Results from the hydraulic surge test are shown in Figure 6-6. Composite samples taken just before

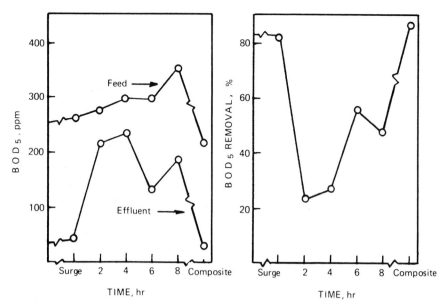

FIGURE 6-6. RBC performance during hydraulic surge. (From Bennett, D., Needham, T., and Summer, R., *Tappi*, 56, 50, 1973. With permission.)

the hydraulic surge indicate the influent BOD to be 270 mg/l and an effluent to be 50 mg/l. At time 0 the wastewater flow was increased by a factor of 4 and continued for a period of 4 hr. After 4 hr, the flow was set back to its original value. During the hydraulic surge, the influent BOD gradually increased to approximately 300 mg/l. At the same time the effluent BOD increased to about 230 mg/l. BOD removal decreased from a level in excess of 80% reduction to approximately 30% reduction during the hydraulic surge. Composite samples taken after the flow was set back to its original level show that the effluent BOD concentration and percent BOD removal rapidly returned to their previous levels.

Performance with an organic slug is shown in Figure 6-7. This was tested by raising the influent concentration from about 230 mg/l to almost 900 mg/l and maintaining this level for 4 hr. The effluent BOD increased to 600 mg/l and the BOD removal decreased to approximately 30%. After 4 hr, the influent strength was returned to its original level, and the effluent concentration and percent removal both returned to their original levels within a few hours.

These two tests discussed above demonstrate the ability of the rotating contactor process to withstand upsets from hydraulic and organic shock loads. Because the vast majority of the active

culture in the system is attached to the rotating media, and because the process does not depend upon sludge recycle for efficient operation, no loss of culture occurs under shock load conditions; return to normal operation is rapid and complete.

Even though the effluent concentration increased and the percent BOD removal decreased during the shock load tests, the actual amount of BOD removal increased. This is shown in Figure 6-8. During both the organic and hydraulic shock load tests, the cultures in the individual stages of treatment were exposed to higher BOD concentrations. They responded by removing BOD at a greater rate. This resulted in a larger amount of BOD removal during the shock load period than would have been achieved during normal operation.

Very similar tests have been conducted by Gillespie et al.[3] on bleached kraft waste, which showed virtually identical results.

GENERAL DESIGN CRITERIA

Most of the process development work done on the rotating contactor process for industrial waste treatment over the past several years has utilized synthetic dairy waste. Several of the following discussions utilize test data generated from the experimental systems described in Figure 6-1 and

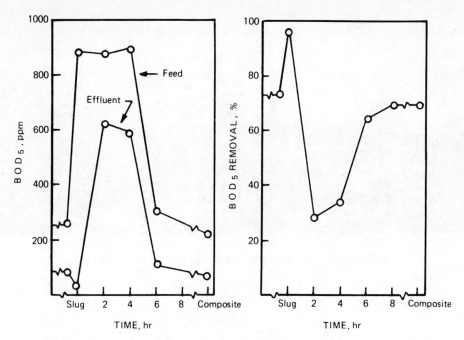

FIGURE 6-7.   RBC performance during organic slug. (From Bennett, D., Needham, T., and Summer, R., *Tappi,* 56, 50, 1973. With permission.)

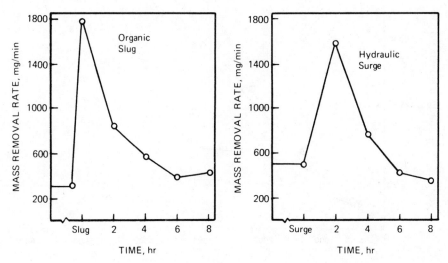

FIGURE 6-8.   Mass BOD$_5$ removal rates during slug and surge experiments. (From Bennett, D., Needham, T., and Summer, R., *Tappi,* 56, 50, 1973. With permission.)

shown in Figure 6-9. The test system shown in Figure 6-9 incorporated pilot plant units with 2-ft-diameter polystyrene discs arranged with either 4 or 8 stages in series and were tested with and without an intermediate clarifier. Synthetic dairy wastewater was supplied to the test units with a system essentially identical to that shown in Figure 6-1.

BOD Removal Kinetics

Figure 6-10 shows BOD removal as a function of applied BOD concentration for three different hydraulic loadings. Data at the lowest hydraulic loading show a linear relationship between BOD removed and BOD applied for BOD concentrations well beyond 4,000 mg/l, which indicates that the process is exhibiting first order behavior within

FIGURE 6-9.   Pilot plant RBC unit for industrial waste treatment.

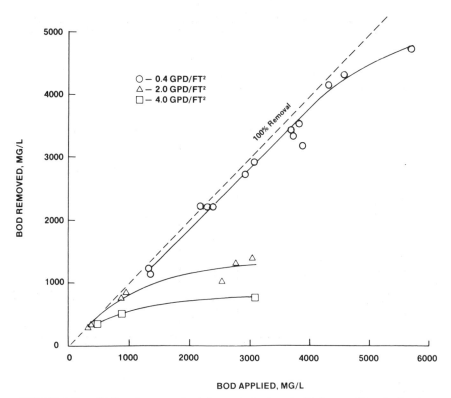

FIGURE 6-10.   RBC performance for dairy waste treatment with intermediate clarification.

that concentration range, because a given percentage of the BOD is removed, independent of the influent BOD concentration. For the two higher hydraulic loadings, a linear relationship extends over a much shorter BOD concentration range. As the trend lines for these two higher loadings begin to depart from linearity, the process departs from first-order behavior. At still higher BOD concentrations, the trend lines begin to approach a horizontal position, which indicates that the process is now approaching zero order behavior. At this point, further increases in BOD concentration will not result in additional BOD removal.

First-order behavior is possible only when there is a relative excess of microorganisms in the system and when oxygen supply is not a limitation. At low hydraulic loadings, the amount of attached culture and the oxygen supply are both in relative excess compared to the BOD concentration; therefore, first-order behavior is exhibited over a wide BOD concentration range. At higher hydraulic loadings, the amount of culture and aeration capacity are no longer in excess, so that first-order behavior is exhibited over a more narrow range of BOD concentrations.

The interrelationship between the attached biomass and the substrate concentration can be seen in Figure 6-11 taken from Welch.[4] It shows the amount of volatile solids attached to the rotating contactor as a function of the substrate concentration in the mixed liquor. Starting at a low substrate concentration, any increase in concentration results in a proportional increase in volatile solids. In the operation of the process, this results in a proportional increase in the amount of substrate removed and first-order behavior is observed. Further increases in substrate concentration result in less than a proportional increase in attached biomass. This results in less than a proportional increase in BOD removal and a departure from first-order behavior is noticed. At still higher substrate concentrations, no additional biomass develops on the surfaces so that no additional BOD reduction occurs. At this point,

zero order behavior is exhibited. Continued increases in the amount of attached culture are not possible, because of the inability of the culture to support its own weight beyond a limiting thickness. Also, the rotational velocity of the contactor exerts a shearing force, which strips off all biomass beyond a limiting thickness.

The other factor which affects BOD removal kinetics in the rotating contactor is oxygen supply. Figure 6-12 (also taken from Welch[4]) shows the effect of increasing rotational speed, and therefore oxygen supply on substrate removal. As the COD concentration was increased, additional COD removal was achieved by increasing rotational speed. This is true, however, only for the high substrate loading rates shown in Figure 6-12. At lower loading rates, where higher degrees of treatment are achieved, increasing rotational speed has not been shown to improve treatment efficiency.[2]

When designing at high loading rates and low degrees of treatment, increasing rotational speed to improve treatment efficiency is not a practical alternative, because power consumption increases exponentially with increases in rotational speed (see Figure 3-9). The significantly increased operating costs do not justify the increased speed. Also, the increased stresses on the rotating con-

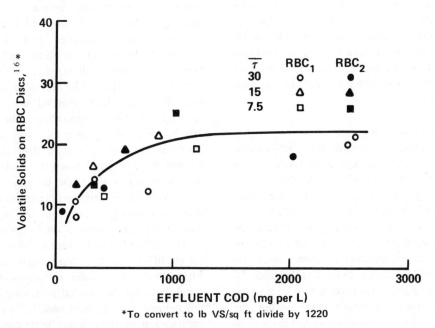

FIGURE 6-11.   Active biomass on RBC units. (From Welch, F. M., *Water Wastes Eng. Ind.*, July/August, 1969. With permission.)

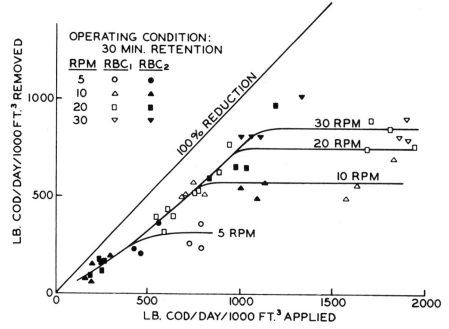

FIGURE 6-12.   Effect of rotational velocity on COD removal.

tactor would require a much more expensive design and construction. It is generally less costly to install additional rotating contactor surface area than to attempt to increase rotational speed for this type of application.

As an alternative to increased speed, it is possible to increase oxygen supply by installing diffusers in the rotating contactor tankage or by operating the rotating contactor within an enclosed oxygen enriched atmosphere. This would be especially beneficial when treating certain types of industrial wastes which tend to be odorous when subjected to biological treatment at high loading rates.

Intermediate Clarification

When treating a concentrated wastewater, excess biological solids will accumulate in the mixed liquor as the wastewater flows through the successive stages of treatment. As these solids reach a high concentration, they suppress the mixed liquor dissolved oxygen concentration and reduce the treatment efficiency in the subsequent stages. Continuous agitation of these solids also decreases their settleability in the final clarifier. To avoid these problems, it is necessary to remove the accumulated solids at some point within the rotating contactor treatment process. This is done by operating an intermediate clarifier after the

first or second stage of a rotating contactor system containing a total of four stages. Since it is not necessary to remove all of the biological solids, the intermediate clarifier can be operated at a relatively high overflow rate, i.e., 1,000 to 2,000 gpd/ft².

The effect of intermediate clarification when treating a concentrated waste at a low hydraulic loading can be seen in Figure 6-13. Operation with the intermediate clarifier enabled first-order behavior to be maintained up to 4,000 mg/l BOD. Without an intermediate clarifier, the process departed from first-order behavior above 1,500 mg/l BOD and BOD removal was significantly reduced at higher concentrations. The same effect can be seen in Figure 6-14. With intermediate clarification, there is only a slight reduction in percent BOD removal when the BOD concentration increases from 2,000 to more than 4,000 mg/l. Without the intermediate clarifier, however, there is a steady decrease in the percent BOD removal for concentrations above 1,500 mg/l.

The effect of intermediate clarification for lower BOD concentration ranges is shown in Figure 6-15. The shape and disposition of the trend lines for the several hydraulic loadings do not indicate any apparent benefit for intermediate clarification at these lower BOD concentrations. Therefore, when treating wastewater strengths

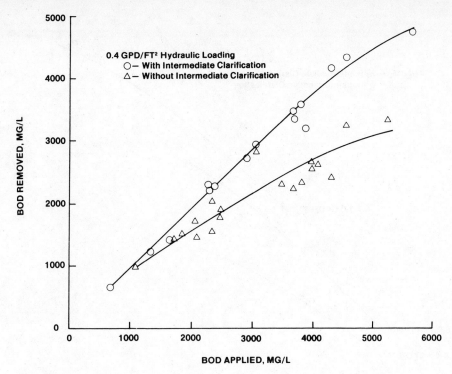

FIGURE 6-13. Improvement of RBC dairy waste treatment by intermediate clarification.

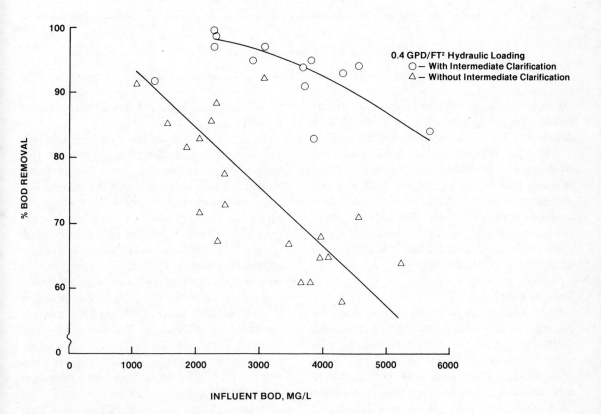

FIGURE 6-14. Effect of intermediate clarification on RBC dairy waste treatment.

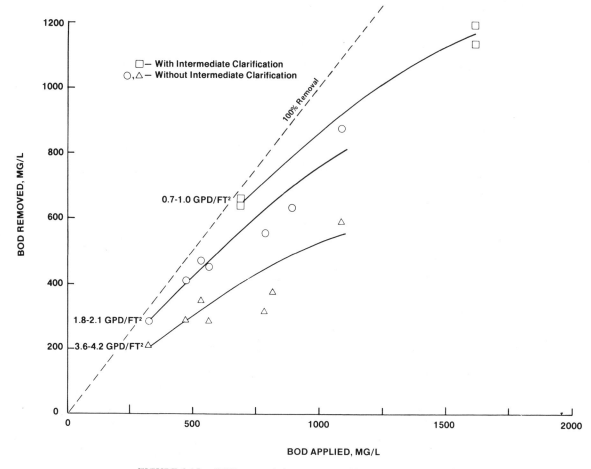

FIGURE 6-15.   BOD removals in treatment of low BOD dairy wastes.

generally in excess of 1,500 mg/l intermediate clarification is recommended especially if it is desired to achieve high degrees of treatment.

The effect of intermediate clarification on the treatment of a yeast manufacturing waste is shown in Table 6-1. Comparing the test data with and without intermediate clarification of comparable loadings indicates that 10 to 20% additional BOD removal can be achieved or operation at a significantly higher hydraulic loading is possible when utilizing an intermediate clarifier.

Enlarged First Stage

It is desirable to operate the rotating contactor process with an enlarged first stage, i.e., where the first stage is larger than the subsequent stages on some wastewater treatment applications. This is done to avoid overloading the first stage and to help attenuate variations in wastewater characteristics. Conditions under which an enlarged first stage is expected to improve process performance

are difficult to estimate and are best determined in a pilot plant study.

Table 6-2 shows some test data from two pulp and paper waste studies with an enlarged first stage. Test data on the more concentrated insulating board waste show that the enlarged first stage permitted higher hydraulic loadings for equivalent levels of treatment. However, for the more dilute kraft waste, the enlarged first stage appeared to yield no significant benefit. This implies that influent BOD concentrations may be an important factor in predicting improved performance from an enlarged first stage. Recent testing on another board waste of 400 mg/l BOD showed no improvement with an enlarged first stage, however. These tests had a lagoon as part of the pretreatment system which suggests that attenuation of fluctuations in wastewater characteristics may also be a factor.

Because the effects of an enlarged first stage will not always be known without a pilot plant

TABLE 6-1

Effect of Intermediate Clarification

Yeast manufacturing waste

With BOD			Without BOD		
gpd/ft²	mg/l	% Reduction	gpd/ft²	mg/l	% Reduction
0.13	1,000	97	0.16	2,040	73
0.22	1,300	95.5	0.32	1,644	80
0.32	1,380	90.6			
0.64	3,670	70.5			

TABLE 6-2

Effect of Enlarged First Stage

Influent BOD	% BOD removal	gpd/ft²
	Insulating board waste	
	Equal sizes stages	
625	74	1.0
565	90	0.75
775	61	0.75
478	87	0.75
	Enlarged first stage	
864	67	2.0
695	91	1.3
	Bleached kraft waste	
	Equal size stages	
212	88	1.9
	Enlarged first stage	
216	91	1.9

study, it may be important to include the flexibility of this operation in the construction of a full-scale plant. Fortunately, this is done easily, because an enlarged first stage can be created simply by removal of the bulkhead or baffle between the first and second stages of treatment.

Wastewater Temperature

Low wastewater temperatures will affect the efficiency of industrial wastewater treatment just as it does for domestic wastewater. Fortunately, there are few industrial wastewater applications where temperatures will be below 55°F. For those cases where low temperatures are encountered, the temperature correction factors shown in Chapter 4 for domestic waste can be utilized for design.

High wastewater temperatures can be a problem for industrial wastewater treatment because of reduced oxygen solubility and the possibility of developing undesirable microbial species. Wastewater temperatures as high as 110°F can be treated with no loss of efficiency at hydraulic loadings of 1.5 gpd/ft² and below. At these loading rates, the rotating contactor can sufficiently cool the wastewater to avoid any deleterious effects on biological treatment. At loadings above 1.5 gpd/ft² wastewater temperatures above 100°F can cause a decrease in treatment efficiency principally through reduced settleability of the biological solids. Table 6-3 shows some general guidelines for wastewater temperature and allowable design hydraulic loading.

Pilot Plant Testing

For many industrial waste applications, specific design criteria should be developed for accurate and economical full-scale design and regulatory agency approval. Several rotating contactor manufacturers make pilot plants available in order to meet these requirements.

Pilot plant studies also demonstrate the operating characteristics, reliability, and effectiveness of a process in treating a particular wastewater, and they will show the compatibility of the equipment with existing facilities.

Several sizes of rotating contactor pilot plants are available for use in test programs. Units of 6 to 10 ft diameter are capable of treating wastewater flows in the 10 to 50 gpm range, and smaller units of 2 to 4 ft diameter can operate in the 0.2 to 10 gpm range, depending on the waste characteristics and treatment requirements.

For applications where it is only necessary to confirm an estimated design loading, the smaller pilot plants are adequate. For application to large treatment facilities, and when it is also important

TABLE 6-3

Allowable RBC Loadings at Elevated Temperatures

Inlet wastewater temperature, °F	Maximum* design hydraulic loading, gpd/ft²
120	1.0
110	1.5
100	2.5
90	3.0

*Without some loss of treatment efficiency.

to more exactly determine pretreatment requirements and the settleability and dewatering characteristics of the biological solids produced, the larger units are recommended.

SPECIFIC DESIGN CRITERIA

Dairy Waste

Data from various pilot plant tests on dairy waste treatment are plotted in Figures 6-16 and 6-17. Figure 6-16 contains test data in lower BOD concentration ranges and for operation without intermediate clarification. For a specific application, all that is necessary is to select the appropriate hydraulic loading for the required percent BOD removal. For influent BOD strengths other than those shown in Figure 6-16, interpolation between the values shown will be necessary. The design relationship for BOD concentrations of 1,600 to 2,200 mg/l differs significantly from those for the lower strengths. This is the concentration at which intermediate clarification becomes beneficial for dairy waste treatment.

Figure 6-17 shows similar design relationships for high strength dairy waste. The data for BOD removals above 85% are from operation with an intermediate clarifier, which was always placed at the midpoint of the rotating contactor treatment process. At very low hydraulic loading rates, the design relationships begin to merge, indicating that the process is approaching first-order kinetics at these loadings. For higher hydraulic loadings and lower degrees of treatment, the process begins to depart from first-order kinetics and at still higher loading rates begins to approach zero order.

For some applications, it may be more desirable to forego the use of an intermediate clarifier and increase the size of the rotating contactor process.

For example, a small wastewater treatment application which would require a single shaft of rotating contactor media while using an intermediate clarifier may only require an increase in rotating contactor size or the addition of a second shaft of rotating contactor media to meet the treatment requirements. Eliminating the intermediate clarifier in this manner simplifies the operation of the plant and, in some cases, may reduce its construction costs. A decision regarding use of an intermediate clarifier should be based on overall cost vs. benefits.

Meat and Other Food-processing Wastes

Figure 6-18 shows design relationships for meat and several other food processing wastes over a wide range of BOD strengths. The data indicate that these wastes differ significantly in treatability. The meat, and poultry and fish-processing wastes show a high level of treatability, because they contain a relatively high fraction of suspended and colloidal matter. The suspended matter is very rapidly adsorbed and flocculated in the biological treatment process. In contrast, the yeast manufacturing and synthetic potato wastes contain high fractions of soluble sugar and unhydrolyzed starch, which are absorbed much more slowly. Data on the yeast-manufacturing waste were obtained with intermediate clarification. The other wastes did not have intermediate clarification.

The difference in treatability among these food-processing wastes points out the importance of conducting pilot plant tests to determine full-scale design requirements.

Beverage Wastes

Test data from treatment of winery and distillery wastes are shown in Figure 6-19. The winery waste data are from pilot plant testing while the distillery waste data are from the full-scale plant described in Chapter 8. Note that although the distillery waste was about twice as concentrated as the winery waste, it was not necessary to operate at half the hydraulic loading for a specific BOD removal. If the distillery waste had been treated at the hydraulic loading of 0.5 gpd/ft², it is likely that it would have showed a percent BOD removal about the same as that for the winery waste at the same loading.

Pulp and Paper Waste

Many pilot plant tests have been conducted over

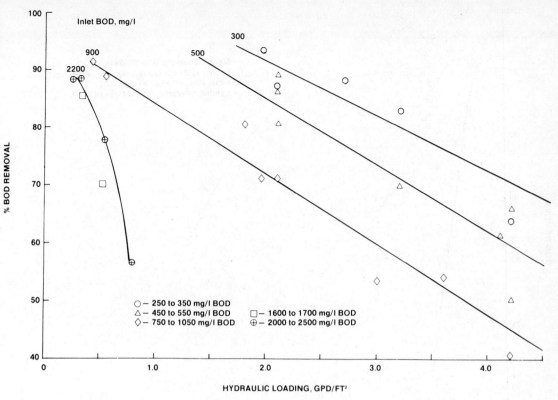

FIGURE 6-16.   Design criteria for dairy waste treatment without intermediate clarification.

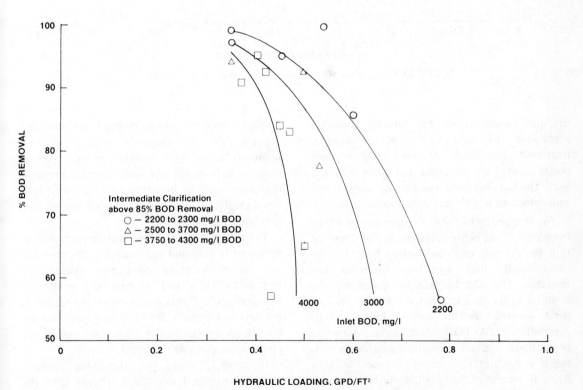

FIGURE 6-17.   Design criteria for high strength dairy waste treatment.

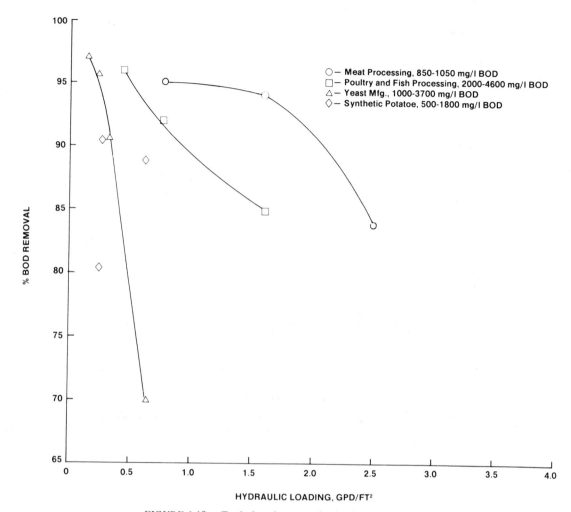

○— Meat Processing, 850-1050 mg/l BOD
□— Poultry and Fish Processing, 2000-4600 mg/l BOD
△— Yeast Mfg., 1000-3700 mg/l BOD
◇— Synthetic Potatoe, 500-1800 mg/l BOD

FIGURE 6-18. Typical performance for food-processing wastes.

the past several years for rotating contactor application to pulp and paper wastewater treatment. The results of some of these pilot plants studied are presented in Figures 6-20 and 6-21. The test data cover a wide range of types and concentrations of pulp and paper waste.

The treatability of pulp and paper wastes differs from that of the other industrial wastes discussed thus far. As pulp and paper wastes become more concentrated, they appear to become less treatable. This can be seen by comparing the required hydraulic loadings for a specific level of BOD removal for the various influent BOD concentrations. As BOD concentration increases, a proportional reduction in hydraulic loading is required. For further increases in concentration to levels in excess of 500 mg/l, there is sometimes more than a proportional reduction in hydraulic

loading. This is thought to be due to an increasing concentration of materials which may be inhibitory to biological activity. Chemicals such as resin acids from the raw materials and inorganic chemicals used in bleaching of pulp and paper could exhibit increasing levels of inhibition as their concentration increases.

The presence of inhibitory substances as well as variations in pulp and paper-making processes and raw materials cause significant changes in treatability from mill to mill and within an individual mill. This is evident from the scatter in the data in Figures 6-20 and 21 and points out the importance of conducting pilot plant studies for pulp and paper-waste treatment.

Figure 6-22 shows the relationship between effluent suspended solids and effluent BOD for pulp and paper wastewater treatment. The

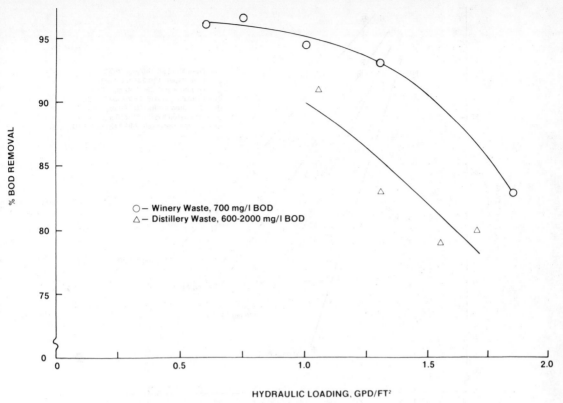

FIGURE 6-19.   Typical performance for beverage waste treatment.

relationship is similar to that found for municipal wastewater treatment, i.e., effluent suspended solids levels are approximately equal to the effluent BOD concentration. For effluent BOD concentrations at or below 40 mg/l, Figure 6-23 shows that approximately half of the remaining BOD is soluble and half is in suspension. Tertiary filtration could then be expected to achieve about 50% additional BOD removal.

The settleability of biological solids produced by the rotating contactor when treating pulp and paper waste was determined in a series of settling column tests. Results of these tests are shown in Figure 6-24. The tests were conducted over a wide range of mixed liquor solids concentrations and for two different types of pulp and paper waste. The data indicate that the percent solids separation at a given overflow rate generally increases as the mixed liquor solids level increases. However, for a specific overflow rate the final solids concentrations are approximately the same for all initial solids levels.

An important characteristic of the trend lines in Figure 6-24 is their relative flatness which indicates the stability of the solids separation process. For example, if a clarifier were designed for 90% solids separation of 150 mg/l at an overflow rate of 1,500 gpd/ft², and the plant suddenly experienced a twofold increase in flow, the solids separation efficiency would only decrease to approximately 85%. This means that the effluent suspended solids concentration would only increase by about 10 mg/l.

The overflow rates indicated in Figure 6-24 should be divided by a factor of 1.5 for use in full-scale clarifier design. This is done to adjust for entrance and exit effects, action of the mechanical solids collector, wind, and thermal gradients in the full-scale clarifier, and for wall effects in the settling column.

Sludge production by the rotating contactor process is shown in Table 6-4 as reported by Gillespie et al.[3] For a variety of pulp and paper wastes the sludge production was about 0.3 to 0.4 lb of volatile solids per pound of BOD removed. Because the biological solids are approximately 80% volatile, the total sludge production is 0.4 to 0.5 lb/lb BOD removed.

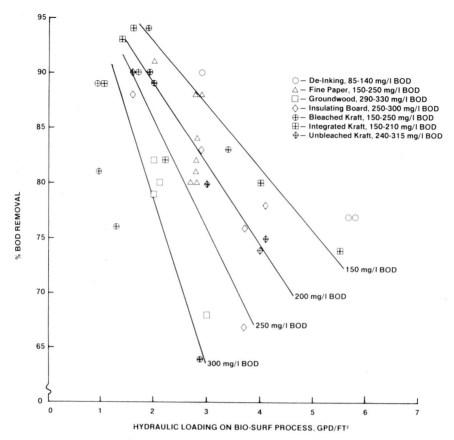

FIGURE 6-20. Treatment of low strength pulp and paper wastes.

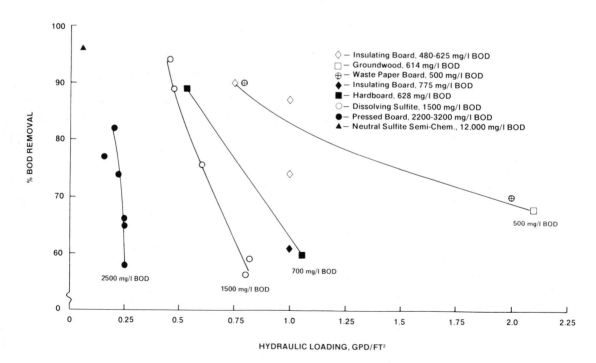

FIGURE 6-21. Treatment of high strength pulp and paper wastes.

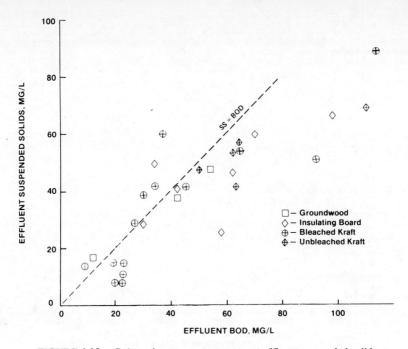

FIGURE 6-22.   Pulp and paper waste treatment effluent suspended solids.

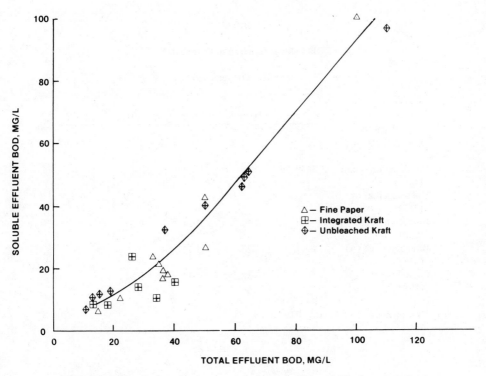

FIGURE 6-23.   Pulp and paper waste treatment effluent BOD characteristics.

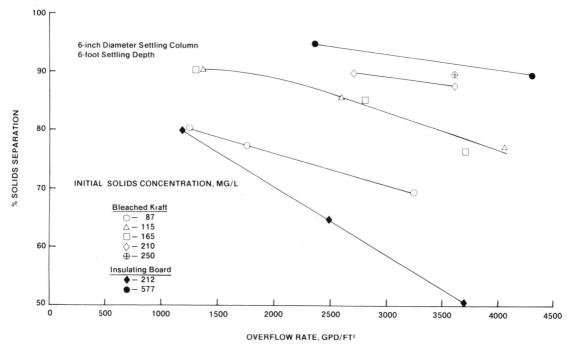

FIGURE 6-24.   Sludge solids settleability pulp and paper waste treatment.

TABLE 6-4

RBC Sludge Generation Rate Data*

#VSS/#BOD removed

Waste influent BOD, mg/l	Hydraulic loading rate, gpd/ft²				
	0–0.9	1.0–1.9	2.0–2.9	3.0–3.9	4.0–4.9
Combination board (1,513)	0.33	0.35	–	–	–
Unbleached kraft (257)	–	–	–	0.39	0.35
Insulation board (240)	–	0.41	0.24	0.31	–
Bleached kraft (A) (235)	0.41	0.27	0.33	–	–
Coated paper (198)	–	0.38	0.34	0.81	–

*Net increase in VSS across RBS unit.

An important aspect of pulp and paper wastewater treatment especially for discharge to Canadian waters is toxicity reduction. Bioassay tests on rotating contactor effluent when treating insulating board mill effluent have been reported by Bennett et al.[2] Test procedures used were those outlined by Sprague, et al.[5] with minor modifications. The Environmental Protection Service, Canadian Department of the Enviroment, has specified rainbow trout (Salmo gairdneri) as a standard test fish for pulp mill effluent. Results from the rainbow trout toxicity test are shown in Table 6-5. Canadian toxicity regulations for the pulp and paper industry specify that a 65% effluent solution be used for bioassay tests. Under specified test conditions, 90% survival for 96 hr is required to pass the test. The data in Table 6-5 indicate that the rotating contactor process can easily meet this requirement.

Similar tests were conducted by Gillespie[3] on several other wastes and the results shown in Table 6-6. The rotating contactor passed the test for all wastes except the sulfite waste at a high loading rate. At a lower loading rate, however, it did pass the test. This is especially noteworthy because sulfite wastes are among the most difficult to treat. Effluents from suspended growth treatment of pulp and paper wastes have not consistently demonstrated toxicity reductions which meet Canadian standards.[6,7]

Refinery Waste

Many pilot plant studies have been conducted with the rotating contactor process at refineries around the U.S. and Canada. BOD removals obtained during several of these studies are shown in Figure 6-25. The wastewater at Refineries A and B and the oily waste at Refinery C were API separator effluents which were treated directly without additional pretreatment. The soluble organic and phenolic wastes from Refinery C and the wastewater from Refinery D received equalization and oil removal prior to rotating contactor process treatment. There does not appear to be any significant difference in rotating contactor performance with these different types of pretreatment. Therefore, API separator effluent can be treated directly without additional pretreatment in many cases.

Several of the data points for Refinery A indicate better performance than other data points. This is thought to be due to fluctuations in wastewater pH values which were generally below 6.5 for the data points at lower hydraulic loadings. Also, there was intermittent use of a strong disinfectant for cooling tower cleaning which often appeared in the wastewater. The wastewater at Refinery D appears to be less treatable than the others. This is because it is a lagoon effluent already receiving partial biological treatment, and its treatability has been reduced. The dashed line in Figure 6-25 can be used for designing a rotating

contactor to treat partially treated refinery waste.

Oil and grease removal is shown in Figure 6-26. Apparently, high oil and grease concentrations are more readily treated than low concentrations, although in both cases, approximately the same effluent concentration is produced. The rotating contactor process appears to handle oil and grease levels as high as 375 mg/l with no deleterious effects. To produce effluent oil and grease levels less than 10 mg/l within the range of hydraulic

TABLE 6-5

Rainbow Trout Bioassay Results for RBC Treatment of Paper Mill Effluents

Test No. 1

Disk speed, rpm	17	
Flow rate, gal/min	0.65	
Untreated effluent	65	No fish alive after 48 hr
concentration, %	38	100% survival after 96 hr
Treated effluent	100	70% survival after 96 hr
concentration, %	65	90% survival after 96 hr
	65[a]	100% survival after 96 hr

Test No. 2

Disk speed, rpm	13	
Flow rate, gal/min	0.5	
Untreated effluent	65[a]	70% survival after 48 hr
concentration, %		No fish alive after 72 hr
	42	80% survival after 96 hr
	10	90% survival after 96 hr
Treated effluent	100	90% survival after 96 hr
concentration, %	65	100% survival after 96 hr
	65[a]	100% survival after 96 hr
Control		90% survival after 96 hr

[a]Indicates replicate tests.

From Bennett, D. et al., *Tappi*, 56, 50, 1973. With permission.

TABLE 6-6

Average Percent Survival After 96 hr in 65% Concentration

Waste	Untreated	Treated
Boardmill	0	98
Kraft	0	95
Sulfite (Long detention time)	0	100
Sulfite (Short detention time)	0	65

From Gillespie, W. J. et al., *Tappi*, 57(9), 112, 1974.

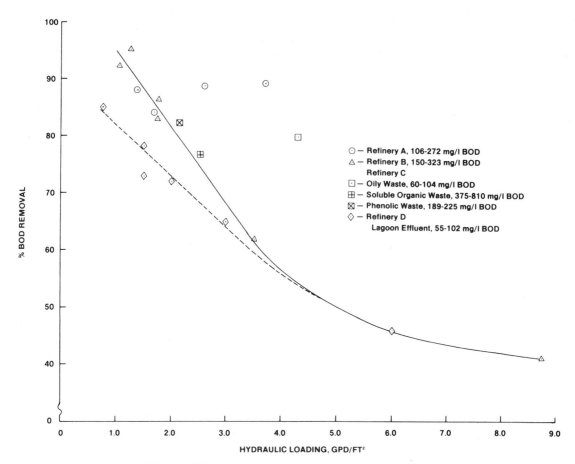

FIGURE 6-25. Refinery wastewater treatment -- BOD removal.

loadings shown in Figure 6-26, it will be necessary to provide additional pretreatment in the form of air flotation to reduce the influent concentration to the RBC process.

Effluent suspended solids concentration is approximately equal to or lower than the effluent BOD concentration for refinery waste just as has been found on other types of wastes. The data points in Figure 6-27 which do not quite conform to this relationship are several of those from Refineries A and D. At Refinery A this probably was due to the low and variable pH levels and the occasional presence of toxic materials. At Refinery D, this was due to the presence of biological solids from the lagoon which preceded the rotating contactor process. Highly dispersed activated sludge solids which did not settle in the lagoon treatment system passed through the rotating contactor process with little change in settling characteristics.

Of all the refinery wastes tested, stable nitrifica-tion was achieved only at Refinery B. This is believed to be due to the low and fluctuating pH levels and the occasional presence of toxic materials at the other refineries. A comparison of nitrification at Refinery B with domestic waste is shown in Figure 6-28. The dashed line in Figure 6-28 is a reproduction of the correlation in Figure 3-40. For domestic wastewater treatment, nitrifi-cation does not begin until a BOD of approxi-mately 30 mg/l is reached. The amount of ammonia nitrogen removal achieved at that point is principally due to cell synthesis but is small because domestic waste contains a relatively large amount of organic nitrogen. Organic nitrogen is readily available as a nitrogen source for synthesis and relatively little of the ammonia nitrogen is consumed. In a refinery waste there is less organic nitrogen present; therefore, the culture consumes a greater portion of the ammonia nitrogen for cell synthesis. At Refinery B, this amounted to almost 50% of the ammonia present. This is important for

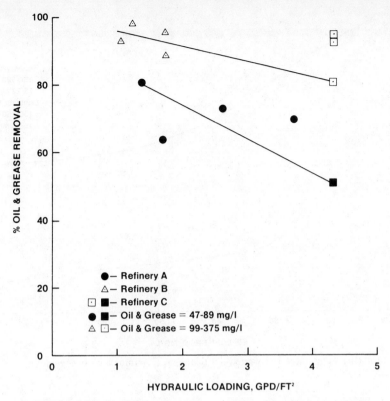

FIGURE 6-26.   Refinery wastewater treatment – Oil and grease removal.

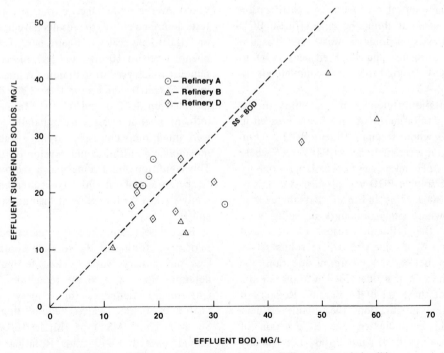

FIGURE 6-27.   Refinery wastewater treatment – Effluent suspended solids.

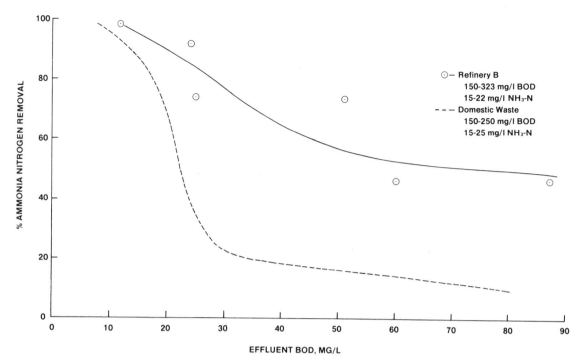

FIGURE 6-28.   Refinery wastewater treatment — Ammonia nitrogen removal.

refinery applications when only moderate amounts of ammonia removal are necessary. Nitrification appears to begin at about the same effluent BOD concentration as on domestic waste, and it is also necessary to reduce the final effluent BOD to approximately 10 mg/l to achieve complete nitrification.

Some of the differences in treatability of the wastewater at Refineries A and B can be seen in Figure 6-29 where soluble effluent BOD is compared to total effluent BOD. At Refinery A where fluctuating pH levels and occasional toxicity existed, the soluble BOD was often half or less of the total BOD. This indicates that there were occasions when larger amounts of solids were present in the effluent because of decreased settleability. At Refinery B where solids settleability was better and where nitrification was achieved, the soluble effluent BOD was often more than half of the total BOD. Because Refinery B was achieving nitrification, the soluble effluent BOD would be increased due to exertion of nitrogenous oxygen demand during the five-day incubation period of the BOD test. If a nitrification inhibitor had been added to the BOD dilution water, the soluble BOD levels probably would have been 25 to 50% lower.

Because influent and effluent phenol levels varied over a wide range during the pilot plant tests described above, the results have been plotted on logarithmic scales in Figure 6-30. For influent phenol levels of 10 mg/l and less, more than 95% removal is achieved at hydraulic loadings generally under 2 gpd/ft^2, and more than 80% removal for loadings up to 3.0 gpd/ft^2. At Refinery C the influent phenol levels were generally higher and were much more consistent so that a stable phenol metabolizing culture could develop on the media. This condition and a somewhat higher wastewater temperature (about 90°F) resulted in phenol removals in excess of 99% at loading rates over 2 gpd/ft^2.

B.C. Research has recently published work on pilot plant testing of the rotating contactor process on refinery waste.[8] Their testing included determination of toxicity reduction. Canadian Government guidelines for refinery waste discharge to marine waters require that Medium Survivor Time (MST) of fish in 100% effluent should exceed 1,440 min. Rotating contactor effluent tested without secondary clarification showed a range in MST from 1,800 min to more than 5,800 min which easily meets the Canadian Government standard.

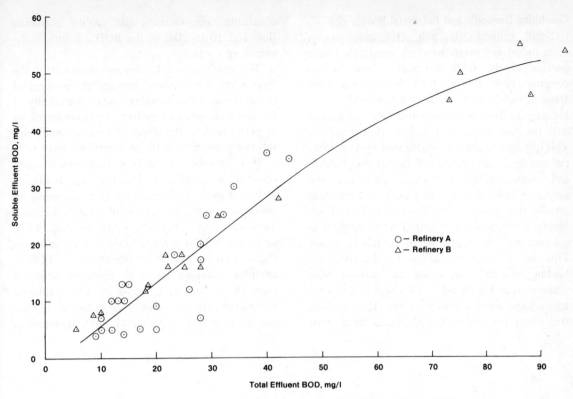

FIGURE 6-29. , Refinery wastewater treatment -- Soluble effluent BOD.

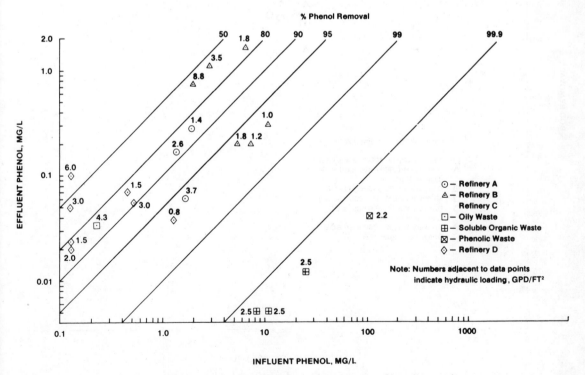

FIGURE 6-30.   Refinery wastewater treatment – Phenol removal.

Combined Domestic and Industrial Wastes

Small municipalities will often have one or more industrial operations which constitute a large portion of the total wastewater flow for the community. In this situation the question often arises as to whether it is more economical for the industry to provide its own treatment or to join with the municipality in a single treatment facility. This question is always complicated by the political and legal ramifications of capital cost funding and responsibility for plant operation and operating costs. The data in Figure 6-31 may help resolve this question for some situations. Combining domestic with industrial waste appears to enhance the treatability of the industrial waste. This can be seen by comparing the hydraulic loading required for a specific percent BOD removal from Figure 6-31 with those required for treating the industrial wastes alone. The enhanced treatability is based on having the domestic waste

constitute approximately half of the hydraulic flow and 10 to 20% of the BOD content of the combined waste.

The exact nature of the enhancement of the treatability is not known. Perhaps, the presence of the domestic waste provides a better availability of inorganic nutrients, or perhaps it buffers variations in pH or reduces the effect of toxic or inhibitory substances inherent in the industrial wastes.

It is possible to achieve ammonia nitrogen removal on combined domestic and industrial wastes. Figure 6-32 contains data pertaining to ammonia removal for combined meat packing and dairy wastes with domestic waste. The dashed line in Figure 6-32 is a reproduction of the curve in Figure 3-40 for domestic waste alone. The BOD to ammonia nitrogen ratio in domestic waste is generally about 10:1, while in the combined wastes this ratio varied between 5 and 10:1. Data on the combined wastes indicate that ammonia

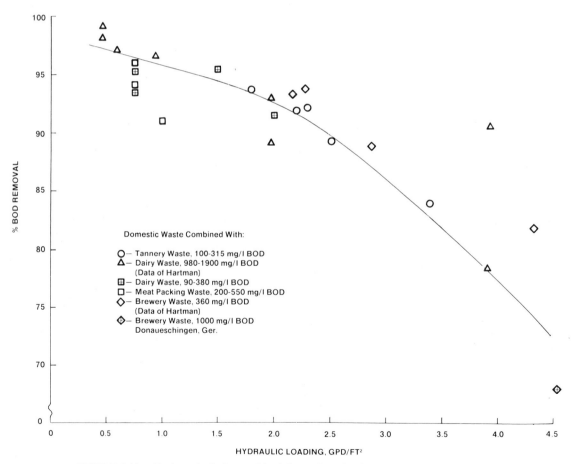

FIGURE 6-31.   Design criteria for combined domestic and industrial wastes – BOD removal.

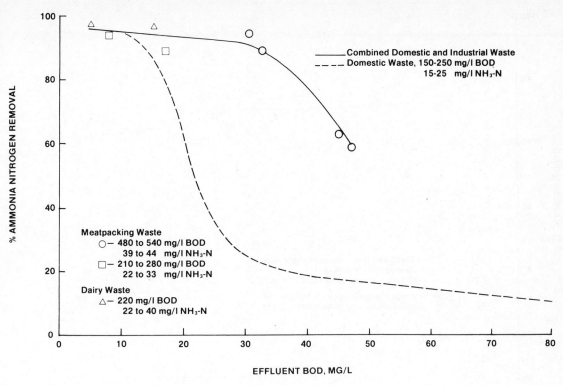

FIGURE 6-32.   Effect of effluent BOD on nitrification.

removal begins at a much higher BOD concentration than it does for domestic waste alone. This results in part from the higher BOD concentration because it exerts a greater requirement for nitrogen for cell synthesis. Because there is a greater concentration of ammonia relative to the BOD strength ammonia oxidizing bacteria can also begin to develop at higher BOD concentrations than when treating domestic waste alone. These two factors appear to affect ammonia removal up to approximately 90% or an effluent of 3 to 5 mg/l.

To produce a final effluent of 1 to 2 mg/l or 95% ammonia removal, however, it is necessary to produce an effluent BOD concentration less than 15 mg/l just as for domestic waste treatment. Similar results have been found by Weng and Molof using synthetic waste.[9] Therefore, to achieve high degrees of ammonia removal on combined domestic and industrial waste, it is necessary to achieve a very high degree of BOD removal. For the meat packing and dairy waste in Figure 6-32, this will require more than 95% BOD reduction.

Miscellaneous Industrial Wastes

Figure 6-33 shows test data from treatment of an animal glue manufacturing waste and a textile waste combined with domestic waste. It shows that as the BOD concentration increases from 200 to 700 mg/l, at approximately the same hydraulic loading, 93% BOD removal is maintained. This indicates that the process was first-order with respect to BOD concentration up to 700 mg/l. Beyond that point, the percent reduction began gradually to decrease, indicating a departure from first-order kinetics and gradually approaching zero-order.

Figure 6-34 shows test data from the animal glue and textile wastes. For various hydraulic loadings, for both wastes, the wastewater pH was much higher than considered optimum for biological treatment. However, because the ph values were very consistent, the biological cultures were able to adapt to these conditions. For degrees of treatment above 60% BOD reduction, the biological activity reduced the pH to a level where it would not have to be further adjusted prior to final discharge to a receiving water.

Olem and Unz[10] have recently published material concerning coal mine drainage treatment with the rotating contactor. At a pH level of about 3.0, they reported very rapid oxidation of ferrous iron by ion oxidizing bacteria attached to the

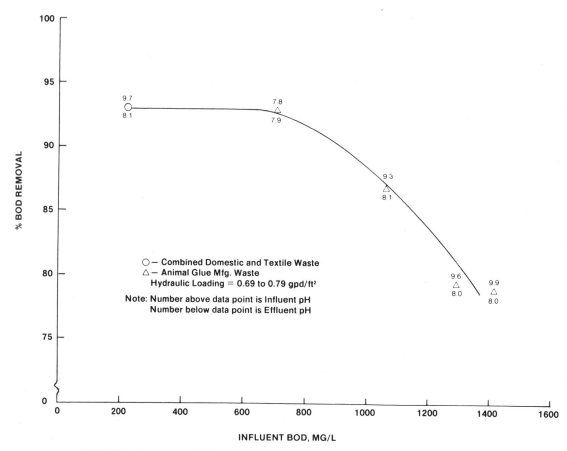

FIGURE 6-33. Effect of influent BOD on treatment efficiency for highly alkaline wastes.

rotating media. Hydraulic loading rates of 7.5 gpd/ft² achieved more than 96% oxidation of up to 300 mg/l of ferrous iron (see Figure 6-35). At 2.7 gpd/ft² ferrous iron effluent levels of 2.0 mg/l were produced. An unexplained side effect of the ferrous iron oxidation is the deposit of ferric compounds on the surface of the plastic media within the attached biomass. These deposits accumulate within the biomass principally on the first stage of media until a biomass thickness of about $\frac{1}{8}$ in. is reached. At this point, the hydraulic shear from media rotation strips off any additional deposits. The accumulation of these deposits has no apparent effect on treatment efficiency, however, their high density increases the weight of the biomass. To reduce the stresses on the rotating contactor equipment, a somewhat smaller media diameter would be employed for mine drainage applications.

The rotating contactor has also been tested on several industrial wastes principally containing ammonia nitrogen. One of these was a wastewater from electronics manufacturing. It received pretreatment for heavy metals removal by chemical addition in a reactor-clarifier before entering the pilot plant. The wastewater contained 15 to 20 mg/l ammonia nitrogen and essentially no organic carbon. At a hydraulic loading of 1.0 gpd/ft² an effluent of less than 1.0 mg/l ammonia nitrogen was produced. The wastewater was treated under somewhat adverse conditions, because alkalinity was low (25 to 30 mg/l), and the pH was low (6.0 to 6.5). Under more favorable conditions, higher loadings are likely to yield equal results. Because the nitrifying culture produced very little waste sludge, there was no need to clarify the rotating contactor effluent to meet effluent suspended solids requirements.

Another high ammonia wastewater treated was from latex polymer manufacturing. This wastewater first received treatment by air flotation and an aerated lagoon. Influent to the rotating contactor unit contained 50 to 60 mg/l ammonia nitrogen, 10 to 15 mg/l organic nitrogen, and

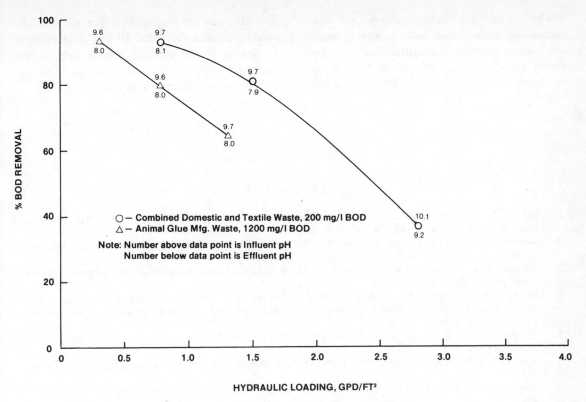

FIGURE 6-34.   RBC treatment efficiency for highly alkaline wastes.

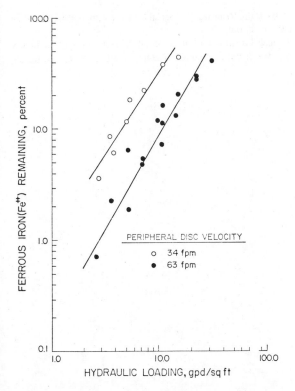

FIGURE 6-35.   Treatment of mine drainage wastewater.

about 10 mg/l BOD. At hydraulic loadings ranging from 0.25 to 0.8 gpd/ft² effluents under 3 mg/l ammonia nitrogen were produced. A nitrogen balance on the process indicated that all the ammonia removed was converted to nitrate. For wastewaters with high ammonia concentrations such as this, it may sometimes be necessary to denitrify following nitrification to avoid discharging high nitrate concentrations to the receiving water. A requirement for denitrification can also be met by the rotating contactor process as discussed in Chapter 4. The process has also been tested on a proprietary chemical processing waste containing about 2,900 mg/l BOD. At a loading rate of 0.12 gpd per square foot, 94% BOD removal was obtained. This was done without intermediate clarification at low pH levels, and with chlorides measuring 4,400 mg/l. The wastewater also had to be chemically detoxified prior to treatment, so this was a rather difficult application.

Finally, base metal mining wastes[11] and a depilatory manufacturing waste have been successfully treated for biological oxidation of thio salts and sulfide ion, respectively.

The rotating contactor process has been successfully applied to a wide variety of wastewaters from proprietary industrial processes. Any wastewater which is not excessively toxic or can be readily detoxified, and contains biodegradable materials can be successfully treated with the rotating contactor process. All that is necessary is to determine the required loading rate for the particular treatment requirements. This, of course, almost always requires a pilot plant study.

REFERENCES

1. **Birks, C. W. and Hynek, R. J.,** Treatment of Cheese Processing Wastes by the Bio-Disc Process, Proc. 26th Purdue Ind. Waste Conf., May 4–6, 1971, W. Lafayette, Ind., 101.
2. **Bennett, D., Needham, T., and Summer, R.,** Pilot application of the rotating biological surface concept for secondary treatment of insulating board mill effluents, *Tappi,* 56, 50, 1973.
3. **Gillespie, W. J. et al.,** A pilot scale evaluation of rotating biological surface treatment of pulp and paper mill wastes, *Tappi,* 57(9), 112, 1974.
4. **Welch, F. M.,** Preliminary Results of a New Approach in the Aerobic Treatment of Highly Concentrated Wastes, paper presented at the 23rd Purdue Ind. Waste Conf., May 7–9, 1968, W. Lafayette, Ind., p. 428; *Water Wastes Wastes Eng. Ind.,* July/August, 1969.
5. **Sprague, J. B. et al.,** *Measurement of Pollutant Toxicity to Fish, Pt. 1–Bioassay Methods for Acute Toxicity,* Pergamon Press, New York, 1969.
6. **Jank, B. E. et al.,** Toxicity removal from kraft bleachery effluent with activated sludge, *Pulp Pap. Can.,* 76(4), 51, 1975.
7. **Harvey, E. and Vaandering, H.,** NSFI pilots Unox system for secondary treatment, *Pulp Pap. Can.,* 76(2), 67, 1975.
8. **Mueller, J. and Bindra, K.,** Rotating disk looks promising for plant wastes, *Oil Gas J.,* Jan. 13, 66, 1975.
9. **Weng, C.-N. and Molof, A. H.,** Nitrification in the biological fixed-film rotation disk system, *J. Water Pollut. Control Fed.,* 46(7), 1674, 1974.
10. **Olem, H. and Unz, R.,** Treatment of Coal Mine Drainage With the Rotating Biological Contactor, paper presented at the 30th Purdue Ind. Waste Conf., May 6–8, 1975 W. Lafayette, Ind.
11. **Jank, B. E., Hamoda, M. F., and LeClair, B. P.,** Biological Treatment of Base Metal Mining Industry Effluents Containing Thio Salts, paper presented at the 48th Annu. Water Pollut. Fed. Conf., Oct. 5–10, 1975, Miami Beach, Fla.

Chapter 7

PROCESS EQUIPMENT CONFIGURATIONS AND PLANT OPERATION

EQUIPMENT CONFIGURATIONS

When the required rotating contactor process surface area for a given treatment application is determined from previous design chapters, the appropriate plant layout can be selected from the figures and discussion in this chapter.

The required amount of active surface area is the primary criterion in process equipment selection. In most cases, several different plant layouts are possible for a given surface area requirement.

A plant layout should be selected based on factors, such as site dimensions available for the installation, site contours, hydraulic head loss through the plant, integration with new or existing auxiliary equipment on the site, and required component redundancy (usually accomplished by parallel flow paths of rotating contactor equipment).

Small Plants

Configuration #1 — Configuration #1 in Figure 7-1 shows four-stage equipment arrangements for surface area requirements up to 300,000 ft^2. Each shaft has open intervals along its length to accommodate cross-tank bulkheads which isolate each section of media as an individual stage to provide the minimum of four stages of treatment for all wastewater flow paths. Wastewater passes from stage to stage through a submerged orifice in each bulkhead to minimize hydraulic head loss. Orifices are sized to provide flow-through velocity sufficient to prevent excessive backmixing. Figure 7-2 shows a four-stage media assembly with the open intervals along its length.

Inlet and effluent connections can be pipes or channels at or below the water level in the tank. The effluent connection or the secondary clarifier weir will establish water level in the contactor tankage. Concrete tanks with common walls and using a trapezoidal contoured bottom are used. This tank has a volume to surface area ratio of 0.12 gal/ft^2.

Configuration #1 has the advantage of parallel treatment flow paths which provide equipment

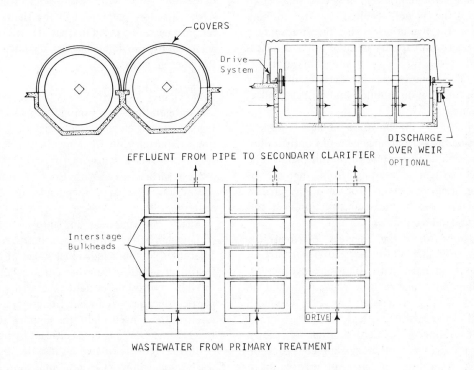

FIGURE 7-1: Plant layout for configuration #1.

FIGURE 7-2: Four-stage media assembly for use in small plant configuration #1.

redundancy. This also permits modular expansion of the plant to higher degrees of treatment and higher flow capacities. A rotating bucket pumping mechanism can also be used if it is necessary to regain several feet of hydraulic head. See Package Plants section in Chapter 4 for a discussion of the operation and use of rotating bucket pumps. These can be supplied by rotating contactor manufacturers and will generally represent about 5% of the cost of the rotating contactor assembly. Additional power consumption is very small compared to the power required by the rotating contactor. Configuration #1 has the disadvantage of less treatment capacity because of the displacement of media by the separating bulkheads for staging. This concept also has extra requirements for influent flow distribution and effluent collection piping.

Figure 7-3 is a drawing for a 0.6-mgd treatment plant using the shaft arrangement of Configuration #1. The unique characteristic of this plant is the compact structure resulting from the use of common wall construction for all of the process components. This reduces construction cost and minimizes land requirements.

Upgrading to higher degrees of treatment requires only that additional shafts of media be installed in parallel with those already present. Upgrading to a higher flow capacity would be done by adding a similar low cost structure adjacent to the existing system.

Configuration #2 — Configuration #2, shown by Figure 7-4 is another system configuration which covers a range of active surface area up to 400,000 ft^2. Here shafts are placed in series and in

several parallel flow paths. Each of the shafts is divided into two stages to achieve the recommended minimum of four stages of treatment for all wastwater flow paths. This configuration has fewer bulkheads and less piping than Configuration #1.

When arranging two or three shafts in series, they need not be placed in a long narrow line of treatment. Instead, they can be placed immediately adjacent to one another (as in Configuration #1) with wastewater following a serpentine flow path through the shafts in series (as in Figure 7-5). Additional flexibility can be achieved by providing by-pass channels between the adjacent shafts.

Large Plants

Configuration #3 — Configuration #3 shown by Figure 7-6 is generally recommended for installations requiring four or more shafts. This system concept covers a range up to 10,000,000 ft^2 of process surface area. With Configuration #3, all wastewater flow is perpendicular to the shaft centerline.

Shafts are installed in simple, flat-bottom, rectangular basins with vertical baffles separating adjacent shafts into individual stages of biological treatment. Baffles can be constructed of any suitable corrosion-resistant material, including plastic, fiberglass, redwood, concrete, etc. Structurally, they need only to support their own weight, since there is no significant hydraulic load on them. Media is rotated against the direction of wastewater flow to prevent short-circuiting beneath the media and baffles. Wastewater flows

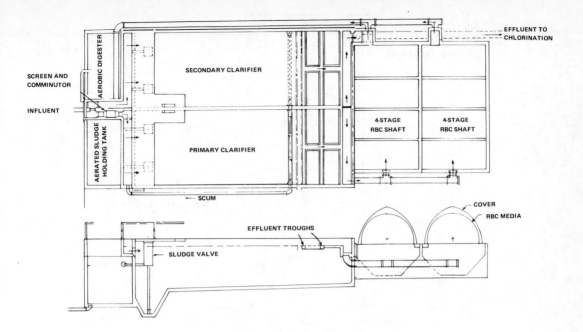

FIGURE 7-3:   Full plant layout drawing for Configuration #1.

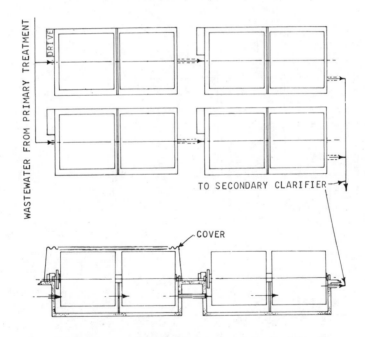

FIGURE 7-4:   Rotating biological contactor Configuration #2.

from one stage to the next beneath the baffles which span the full width of the tanks. The resulting small head loss permits many shafts of media to be installed in series without encountering hydraulic problems.

In cases where more than four shafts of media in series are selected, some interstage baffles can be eliminated provided that at least four approximately equal size stages of treatment are maintained. When using 8 or 12 shafts in each row, baffles would be placed after every second or third shaft to yield four stages of treatment. There would then be two or three shafts per stage. For eight or more shafts per row, staging can be

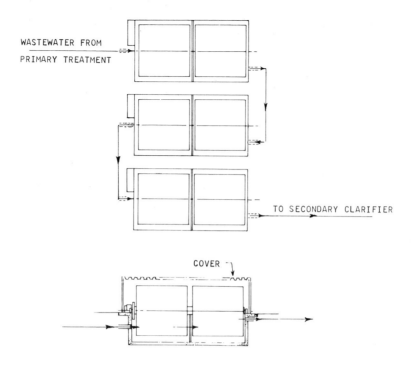

FIGURE 7-5: Alternate flow path for multiple stages.

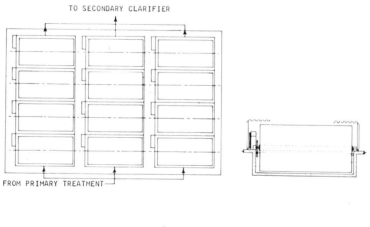

FIGURE 7-6: Rotating biological contactor Configuration #3.

achieved without the need for baffles. This is done by rotating adjacent shafts in opposing directions. The mixing patterns which develop provide an equivalent number of stages equal to one half the number of shafts. A V-notch inlet distribution weir and a flat effluent collection weir are suggested before the first and after the last shaft of media, respectively, for each row of shafts. Fillets are placed at the influent and effluent ends of the tank to prevent solids deposition.

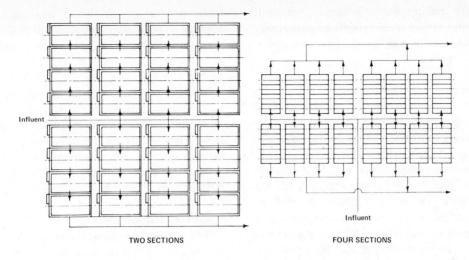

TWO SECTIONS   FOUR SECTIONS

FIGURE 7-7:   Rotating biological contactor Configuration #4.

Configuration #3 maximizes the use of media by eliminating bulkheads and minimizes the amount of concrete in the tankage and walkways as compared to Configurations #1 and 2.

When using a single flow path for Configuration #3, it may seem that component redundancy has not been provided. However, if one shaft assembly in the flow path must be shut down for repairs, it is not necessary to drain the tank, which allows continued operation of the balance of the shafts with only a moderate loss of overall treatment efficiency.

Configuration #4 – Configuration #4 shown by Figure 7-7 is an extension of Configuration #3 to 40,000,000 ft² of active system surface area. This is accomplished by constructing Configuration #3 in either two or four parallel sections. Configuration #4 requires more influent distribution and effluent collection piping and channels than #3, but can reduce head loss and may fit more conveniently into some site plans.

Wastewater is distributed to the various parallel flow paths from a common influent channel. The channel will be of sufficient depth and width to minimize hydraulic head loss and simplify the uniform distribution of the wastewater. The channel will often have to be aerated to keep the wastewater fresh and avoid solids deposition. Connection of the channel to the process flow paths can be done in a variety of ways: V-notch overflow weir, submerged piping with control valve, submerged channels with sluice gates, etc. Any method will be adequate as long as

there is a reasonably uniform distribution of flow across the width of the rotating contactor tankage.

Hydraulic Considerations

Distribution of wastewater flow to and from each rotating contactor with flow either parallel or perpendicular to the shaft is usually of minor significance to process performance. Media rotational velocities and local turbulence are much greater than wastewater flow velocities through the process tankage. Mixing within a stage is very thorough, and the influent-effluent flow distribution makes little difference.

Over the range of normal flows, head loss through the system is almost negligible. In larger, multi-shaft plants, most of the head loss occurs over the influent and effluent weirs and can usually be kept within 6 in. Head loss from the baffles separating adjacent shafts is usually less than 0.2 in. from first to final shaft. In smaller, single shaft, multistage systems with flow parallel to the shaft, the head loss from influent to effluent through the submerged orifices in the interstage bulkheads is normally under 0.5 in.

The rotating contactor process generates biological solids which settle at a comparatively high rate. This is the result of the large floc or particle size. For this reason, it is important that effluent mixed liquor from the process be transported to the final settling tank in a manner which will not physically shear the floc into finer, less settleable particles. Mixed liquor effluent is best transported by gravity flow to the clarifier with as short a fall as possible over weirs, into channels, etc. If pumping is necessary, pump selection becomes important. Screw pumps or low speed centrifugal pumps may be acceptable, while high speed centrifugal pumps should not be used.

Enclosures

The rotating contactor process must be covered or enclosed with a minimal structure. The enclosure serves two main functions. Operation in low ambient temperatures requires a housing to limit wastewater temperature reduction through evaporative heat loss. Rotating surfaces should not receive direct exposure to sunlight to avoid the growth of algae on the outer surfaces, especially for nitrification applications. While the algae will not affect process efficiency, they do represent a nuisance because of additional sludge production and poor settleability.

The rotating surfaces act as an evaporator. When housed in a building, the atmosphere within the building will be at or near 100% relative humidity. A conventional building, therefore, must be heated, at least during the colder months, to prevent continuous condensation, wetting, and accelerated corrosion of the building structure and fixtures.

Building materials must be carefully selected for corrosion resistance and thermal insulation characteristics. Even heated buildings will experience some condensation and water vapor on walls and exposed surfaces during winter operating periods. Acceptable building materials include painted concrete block, precast concrete panels, treated wood, asbestos panels, fiberglass, etc. Operating experience with all such materials has been satisfactory. Painted steel for internal surfaces or structural members should be avoided unless used in conjunction with very effective thermal insulation.

The simplest and generally the most cost-effective enclosure for the rotating contactor process consists of a prefabricated fiberglass or plastic housing. Rotating contactor manufacturers furnish such a housing to cover each individual shaft assembly as shown in Figure 7-8. The shaft drive equipment is usually included within the enclosure and is engineered to withstand the high humidity conditions. The enclosure is provided with access panels and doors to permit direct access to mechanical equipment.

The housing is made and assembled entirely with plastic parts which precludes corrosion

problems. Heating or forced ventilation is not required. The housing provides effective thermal insulation, so that, even in severe winter conditions, wastewater temperature loss is normally held to 1 or 2°F.

Demand for oxygen by the biological process is not a critical criterion for housing design. Covers do not require louvered areas to permit positive air flow. These structures are sufficiently permeable through assembly joints to permit adequate air and oxygen transport to the process. The same applies to more conventional building structures.

Subjacent Clarification

For applications where limited land is available, a configuration where rotating contactor equipment is mounted over a subjacent clarifier may be the most cost-effective alternative. This application was discussed in detail in Chapter 5 as a means of upgrading an existing primary treatment plant, however, it can be equally applicable to new plant construction. The configuration shown in Figure 5-1 takes a structure which normally can provide only clarification and utilizes it for both secondary treatment and clarification. This has some significant cost-saving potential for many applications.

Intermediate Clarification

As discussed in Chapter 6, treatment of concentrated industrial wastewaters can often be achieved

FIGURE 7-8: Typical enclosure for rotating biological contactor.

more efficiently when including a step of intermediate clarification. This can be done in several ways. For Configuration #1 with a single shaft of media, a clarifier would be installed adjacent to the rotating contactor tankage. Mixed liquor would flow by gravity from Stage 2 to the clarifier. Clarifier overflow would then either flow by gravity or be pumped. (depending on head loss through the clarifier) to the third stage of media for continued treatment. For Configuration #2 (and Figure 7-5), the intermediate clarifier would be placed after the first shaft of media, and wastewater flow would be by gravity through the subsequent shaft(s).

Incorporating intermediate clarification into Configuration #3 can be done in either of two ways. One is to replace the baffle between Stages 2 and 3 with a solid wall containing a fillet like that on the influent end of the tank. A weir structure would then be installed along the wall for the full width of the tank. It would collect the Stage 2 effluent and conduct it to the intermediate clarifier located away from the rotating contactor tankage. The overflow from the clarifier is then pumped to Stage 3 where biological treatment continues. This operation allows all the shafts to be installed in the same tank structure.

An alternative approach is to build separate tank structures for the first two stages and the last two stages. The effluent from Stage 2 flows to the intermediate clarifier. Clarifier effluent then flows to a separate tank structure containing Stages 3 and 4. While this approach requires two tank structures for the rotating contactor equipment, it does permit gravity flow of wastewater through the entire treatment plant.

Enlarged First Stage

For domestic and industrial wastewater applications requiring just rough BOD removal at high loading rates, it has sometimes been found desirable to increase the relative size of the first stage of the media to avoid overloading. Overloading of the first stage can cause oxygen-limiting conditions and/or development of undesirable microbial species. Overloading is avoided by increasing the size of the first stage relative to the subsequent stages. This can be done simply and conveniently for all equipment configurations. For Configuration #1, the cross-tank bulkhead separating Stages 1 and 2 can be deleted and half of all the media on the shaft would be acting as

the first stage. Because this configuration would be used only for applications at high loadings and low degrees of treatment, the use of fewer than four stages of treatment would not result in any significant loss of treatment efficiency.

In Configuration #2 (and Figure 7-5), the first shaft of media would be a single stage without cross-tank bulkheads. Subsequent shafts could then be constructed with several stages to provide multistage operation. For Configuration #3, the placement of interstage baffles can be adjusted to provide any desired distribution of media in the various stages of treatment. If the baffles are designed to be removed, adjustments can be made in the field after the plant is in operation to optimize BOD removal.

Package Plants

Equipment configurations for package plant operation were discussed in detail in Chapter 4 principally for applications utilizing a septic tank for pretreatment. Septic tank pretreatment is generally applicable to wastewater flows up to 50,000 gpd. For aerobic pretreatment and for flows up to about 300,000 gpd, the configuration in Figure 7-9 can be employed. Outside tankage construction would generally be of concrete. Internal tankage could be of either steel or concrete with common walls used wherever possible. Pretreatment is achieved with a fine screening device. Screenings fall directly into an aerobic digester. Fine screening devices are in use in several rotating contactor installations, and when there are not excessive amounts of grit or grease in the wastewater, they are satisfactory as the only means of pretreatment.

Screened wastewater flows into an aerated flow equalization tank to dampen the fluctuations often experienced in small plants. After equalization, the wastewater passes through four-stage rotating contactor treatment and then to secondary clarification. The settled solids are pumped into the adjacent aerobic digester. Clarifier effluent is then disinfected and discharged. Periodically, the accumulated solids in the digester are removed for ultimate disposal.

For most climates, the rotating contactor and fine screens would have to be enclosed. For some cold climates and for installations near residential areas, it will be beneficial to enclose the entire facility. Because of its compact construction, completely enclosing the plant should not be too expensive.

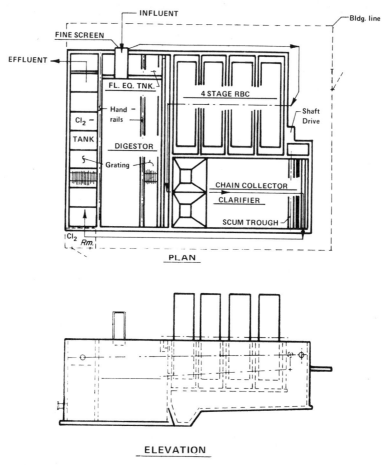

FIGURE 7-9: Package rotating biological contactor plant configuration.

PLANT OPERATION

Treatment Plant Start-up and Normal Operation

To start up a rotating contactor plant, wastewater from the primary treatment system should be passed through the shafts of rotating media and the final settling tank. For the first several days of operation there will be no detectable biological growth on the rotating surfaces, however, aeration and flocculation by the rotating surfaces will result in some BOD removal. After 3 or 4 days of continuous operation, biomass may not be visible but should be noticeable by touching the rotating surfaces; biological treatment will then have started. After a week to ten days, biomass will be visible, and BOD removal should be near that anticipated for steady state operation. At this time, some sloughing of the biomass may occur but will not significantly affect treatment efficiency. After two or three weeks of operation, the biomass will be completely established, and the anticipated level of BOD removal should be achieved. If the treatment plant is also designed to achieve nitrification, an additional two or three weeks will be required for the nitrifying bacteria to become established on the latter stages of rotating media. The beginnings of nitrification can be determined by the presence of nitrite ion (NO_2^-), in the effluent. When nitrite disappears and nitrate ion (NO_3^-) appears complete, nitrification is then underway.

Starting a plant at less than design flow is identical to that for one at full design flow. If there are several independent flow paths of rotating contactor equipment, it may be possible to leave some of the equipment idle and save its operating costs. However, if all the equipment is operated, the degree of treatment achieved will be much higher and the amount of waste sludge produced will decrease. Start-up during winter will be similar. However, if wastewater temperature is less than 55°F, then the various stages of biomass development could be delayed by a factor of two or three times the normal period, depending upon how low the temperature is.

Within the stages which are accomplishing only BOD removal, fully developed biomass will typically appear as a grey-brown, shaggy mass with a thickness of $\frac{1}{16}$ to $\frac{1}{8}$ in. Figure 2-2 shows a typical biomass. The growth in latter stages (which may be achieving nitrification with municipal waste will be much thinner $\frac{1}{20}$ to $\frac{1}{30}$ in.), less shaggy, and more golden brown in appearance. The growth may also be very irregular in thickness due to the activity of predators, such as protozoa, rotifers, and worms. The small pilot plants in Figure 3-26 show this characteristic gradation in the growth.

The dissolved oxygen content of the wastewater beneath the rotating media (mixed liquor) will vary from stage to stage. A plant designed for BOD removal will normally have 0.5 to 1.0 mg/l D.O. (dissolved oxygen) in the first stage. This will gradually increase to more than 1.0 mg/l and often 2 to 3 mg/l in the last stage. A plant also designed for nitrification will have more than 1.0 mg/l D.O. and often 2 to 3 mg/l in the first stage. The last stage will normally have more than 4 mg/l and often 6 to 8 mg/l D.O.

Suspended solids concentration of the treated wastewater leaving the last stage of media will be about one half of the influent BOD concentration for a plant designed for just BOD removal. A plant also designed to achieve nitrification will discharge suspended solids of about one third the influent BOD concentration.

Operating Problems

During operation of a rotating contactor treatment plant, process operating difficulties can arise if unusual waste constituents or adverse conditions are present, or if the plant is subjected to overloading conditions not anticipated in the original design. This section discusses some potential operating difficulties and outlines possible causes and alternative means of correction.

Problem – Excessive sloughing or loss of biological growth from the media.

Cause 1 – Influent waste contains toxic or inhibitory substances (e.g., hexavalent chromium and other heavy metals, chlorine, and other disinfectants).

Solution – Determine what may be responsible; generally it will be an industrial discharge. Determine its concentration, discharge frequency, and duration. Elimination of the toxic substance is the best solution, although, this may not always be possible. In the event that the toxic substance cannot be eliminated, dampen loading peaks and create a uniform concentration of the toxic or inhibitory substance to permit an acclimated culture to develop.

Equalization of the inhibitory substance can be best accomplished at the source. If this is possible, it must be accomplished at the treatment plant. When the corrections must be made at the treatment plant, dampening would be accomplished by an aerated equalization tank.

All biological processes are subject to impairment from significant doses of toxic substances. The rotating contactor process is no exception in this regard, though it is likely to be somewhat less suspectible to short-term slugs of toxicity due to the large inventory of biological solids and short wastewater retention time.

Cause 2 – Severe and unusual variations in influent pH. pH conditions in the range of 6.0 to 8.5 will generally not cause any sloughing problems. However, if severe variations exist consisting of periods of low pH (below 5) or high pH (above 10.5), loss of biomass may result.

Solution – Neutralization of pH can be accomplished by chemical addition and/or equalization. Neutralization is required to insure that influent pH to the system is maintained within the range of 6.5 to 8.5 at all times. Performance will be optimized by maintaining pH within these limits with as little variation as possible. All biological processes exhibit instability under changing pH conditions. Experience has shown the rotating contactor process to be comparatively tolerant of pH variations, while continuing to operate satisfactorily. Again, this is attributed to the large amount of biological solids maintained on the rotating media.

Problem – Development of white biomass. In cases where low O.R.P. (oxidation-reduction potential) exists either in the raw sewage or the rotating contactor system, it is possible to develop organisms on the media which appear white in color. There is no immediate concern if the white organisms (probably *thiotrix* or *beggiatoa*) appear in limited areas on the media. However, if this form of biomass appears to predominate over all the surface, reduced treatment efficiency may be expected. The probable causes of the presence of these organisms and the means by which they can be eliminated are discussed below.

Cause 1 — Influent septic sewage and/or high H_2S concentration. Severely septic sewage and industrial discharges with high H_2S concentrations can cause predominance of a white filamentous growth on the media. This will usually be the sulfur oxidizing bacteria *beggiatoa*.

Solution — This situation can be improved in several ways: (a) preaeration of the influent waste; (b) addition of chemicals to raise the O.R.P. of the waste; (c) eliminate sulfur from the wastewater if it is due to an industrial discharge. The amount of preaeration required will depend upon the O.R.P. of the waste and the pH. If chemicals, such as hydrogen peroxide or sodium nitrate, are added, the dosage will also be based on the O.R.P. of the waste and should be determined by experimentation.

Cause 2 — Overloaded first stage. When severe organic overloads occur on the first stage of the process, due to unexpected increases in flow and BOD in the influent waste, it is possible to develop the white filamentous biomass on Stage 1.

Solution — Provide a larger amount of surface area on the first stage. This may be accomplished by removal of the bulkhead or baffle between Stages 1 and 2. Through this operation, it is possible to reduce the first stage organic loading by a factor of two, thereby eliminating or at least minimizing the overload condition. See the discussion of enlarged first stage earlier in this chapter.

Problem — Decrease in treatment efficiency. The rotating contactor, as a fixed film reactor, is generally less susceptible to upset from shock loadings and hydraulic and organic variations than suspended growth processes. Nonetheless, it is a biological treatment system subject to a lesser extent to all the same factors which can cause reduced efficiency in any biological treatment system. The following paragraphs outline some of

the major factors which can adversely affect process efficiency.

Cause 1 — Reduced wastewater temperature. As wastewater temperature decreases below 55°F, a reduction in biological activity will result, and a corresponding decrease in performance will be observed. The exact amount of performance reduction will depend on the operating load and wastewater temperature at the time. Wastewater temperature is a very critical factor during plant start-up. Development of a culture (especially a nitrifying culture) under very low temperature conditions may require as long as 8 to 12 weeks to achieve steady state conditions with wastewater temperature as low as 45°F.

Cause 2 — Unusual variation in flow and/or organic loading to the process. Because of the large biological culture maintained on the media surfaces, wide variations in wastewater flow rate are effectively treated without loss of attached biomass, and without substantial upset.

For carbonaceous BOD removal sustained periods of very high organic load can temporarily cause an overload of the first stage of a multistage system. Overloading is defined as a BOD load so high that the first stage of the rotating contactor cannot meet the demand for oxygen by the attached growth. While not markedly impairing performance, sustained overload conditions can result in some reduction in BOD removal, and may alter the settling characteristics of the biological solids.

Conditions which result in overloading are not easily predicted. For domestic wastewater application, a sustained first-stage organic load in excess of 20 lb $BOD_5/1,000$ ft^2/day may be considered a potential overload condition. A period of 6 to 8 hr at or above this load could be considered an extended period. Prior knowledge of expected load variations make it possible to design the treatment plant to accommodate them.

The microorganisms which accomplish nitrification respond differently to changing conditions than do carbon oxidizing organisms. Nitrifying organism generation times of 10 to 20 hr compare to approximately 20 min for carbon oxidizing bacteria. Therefore, the time to develop or re-establish a stable nitrifying population is relatively long.

The principal concern with widely varying waste loads to the nitrification process results from the limited ability of nitrifying organisms to compete with the heterotrophic (carbon oxidizing) organisms. Under normal steady state operating conditions, nitrifiers will begin to populate the surfaces after the BOD_5 concentration has been reduced to the range of 20 to 30 mg/l. If a sustained increase in flow or organic load to the carbon oxidizing portion of the system occurs, BOD concentration in the nitrifying stages will increase. Sustained levels in excess of 30 mg/l BOD_5 may eliminate the nitrifying cultures attached to the rotating surfaces. After reduction of the load, it can take several days to reestablish the nitrifying population. Insufficient nitrification will occur during this period. To avoid these conditions, flow equalization may be required.

Widely varying influent ammonia concentrations due to irregular industrial discharges may also require equalization, depending on their magnitude and duration. A peak load of two times the previous daily average ammonia load when sustained for periods longer than 3–6 hr may increase effluent concentration by an amount nearly equal to the incremental increase in influent concentration.

Treatment plants with anaerobic digesters must avoid irregular discharge of large volumes of supernatant back to the plant influent. While this will have little effect on BOD removal efficiency, the high concentrations of ammonia nitrogen may result in high effluent ammonia levels. Supernatant should be returned to the plant influent in a controlled fashion to avoid peak ammonia loads.

In cases where overloading is suspected, samples of the process influent and effluent should be taken as often as is necessary, and analyzed for BOD (total and soluble) suspended solids and ammonia nitrogen. Influent flow to the plant should be noted at all sample times. In most cases where the influent flow and/or organic load peaks are less than twice the daily average over a 24-hr period, little decrease in process efficiency will result. In treatment plants where these hydraulic and/or organic load parameters are exceeded for a sustained period, the above sampling and analysis program will indicate if corrective action is required.

Cause 3 — pH. pH should be maintained between 6.5 and 8.5 for optimum biological efficiency. Exceeding this range will cause a decrease in efficiency until proper conditions are restored. This is especially important for nitrification where sustained pH conditions beyond the range of 6 to 9 can cause a complete loss of nitrification capability.

Alkalinity levels (expressed as $CaCo_3$) should be maintained in at least seven times the influent ammonia concentration in the raw wastewater, in order to allow the nitrification reaction to go to completion without adversely affecting pH conditions.

OTHER POTENTIAL PROBLEM AREAS

Pretreatment

Proper operation of primary treatment facilities is important to optimize performance of the rotating biological contactor process. Removal of settled sludge and surface skimmings from a primary clarifier must be accomplished frequently enough to prevent the development of septic conditions. Evidence of excessive solids accumulation or inadequate removal rate is usually seen in release of gas bubbles at the clarifier surface, and frequently a decrease in pH through the clarifier.

The rotating contactor process will treat grease concentrations up to 200 mg/l (as hexane solubles) with no loss of treatment efficiency. For industrial or combined industrial/municipal wastewater applications, where a higher grease content is anticipated, air floatation or other means should be employed to reduce the grease content to less than 200 mg/l prior to entering the rotating contactor system.

If inadequate grit removal and primary solids reduction occurs, solids may accumulate in the rotating contactor tankage. Accumulated solids will become anaerobic, can result in the development of odors, and may exert a deleterious influence on process performance. The products of anaerobiasis will exert additional oxygen demand in the mixed liquor and may result in the production of additional soluble BOD in the system. One indication of accumulating solids may be an unusual decrease in pH through the process. In the event that such an operational problem develops in an operating system, the tankage should be pumped free of the solids, and the type and concentration of the solids should be determined in order to establish the source.

Final Clarifier Operation

In order to assure producing high quality effluent, it is necessary that the final clarifier be properly designed and operated. Sludge collection and removal must be done at a sufficient rate to prevent the development of septicity. Sludge removal from the clarifier should be performed in a manner which maintains the high settled solids concentration and not unnecessarily dilutes the sludge.

The result of too infrequent or inadequate sludge pumping is the development of anaerobic conditions in the sludge. Anaerobiasis can result in the release of gas bubbles, causing solids floatation and excessive suspended solids and BOD in the effluent. The generation of hydrogen sulfide within the anaerobic sludge can cause odor problems. Release of additional soluble BOD_5 and generation of ammonia is often the result of too infrequent sludge removal. Denitrification of a nitrified effluent due to reduced O.R.P. in the final clarifier sludge causes the release of nitrogen gas and also results in floatation of sludge.

There are no definite rules for sludge handling in the final settling tanks; only field experience with a particular waste will determine the optimum sludge withdrawal technique. Good operating practices will dictate that sludge pumping be accomplished periodically around the clock. A pumping frequency of 4 hr is usually satisfactory. The pumping period must be determined by trial and error with the objective of just removing the sludge, but at the same time not pumping so much as to dilute the solids. This frequency of pumping will usually dictate a system designed to function automatically with a timer.

Plunger or diaphram pumps are a better alternative than telescoping valves or centrifugal pumps as an effective means of sludge pumping without dilution.

Operating Flexibility

A limitation in the operation of the rotating contactor process, which is sometimes mentioned when comparing it to the activated sludge process, is an apparent lack of flexibility. Once a rotating contactor plant is constructed with a specific number of shaft assemblies rotating at a specific speed, little can be done to change the operation of the plant to accommodate changes in operating conditions, while for an activated sludge plant, the rate of aeration and rate of sludge recycle can be adjusted to meet changing conditions. On the surface this seems like a significant difference, however, upon close examination, the difference is not nearly so great, and in many cases, just the opposite could be said.

An increase in aeration rate in the activated sludge process can improve performance only if there is an oxygen limitation. Since this is generally not the case in municipal waste treatment, the only real flexibility is to reduce the aeration rate on an underloaded plant to save operating costs or to vary the rate of aeration along the length of an aeration tank operated with "plug flow" (tapered aeration), again to save operating costs. The same can be done in a rotating contactor plant. In an underloaded plant, one or more of the parallel flow paths of shafts can be shut down to save power costs. For seasonal changes in flow, rotating contactor units can also be taken out of operation. The attached biomass need not be removed but is allowed to completely dry out. When restarted, a new biomass will develop within a few days.

The use of multispeed drive motors also allows independent adjustment of rotational speed in each rotating contactor stage. This is especially true for the air drive system described in Chapter 3.

The other area of flexibility in activated sludge operation is sludge recycle. Superficially, this seems to offer the operator a simple means of process control under changing conditions. However, increasing solids recycle puts additional hydraulic and solids loads on the clarifier. Loss of biological solids into the effluent and/or inability to adequately thicken the return sludge will limit the amount of sludge recycle the system can tolerate. In any case, it is the solids retention time or SRT (MLVSS inventory/sludge wasting rate, including overflow into effluent) that determines BOD removal efficiency and solids settleability. Thus, the plant operator must have considerable experience and relatively sophisticated controls to utilize this flexibility to adjust to changing conditions. This becomes doubly complex when using two steps of aeration and settling, to achieve both BOD removal and nitrification, because it is now necessary to control two different SRT values simultaneously.[1] By contrast, in the rotating contactor process, there is no need for such control. The amount and type of biomass that develops on each stage of media will be that best suited to treating the wastewater as it undergoes a progressively increasing degree of treatment. Because solids are not recycled, the settling characteristics are of no consequence in process performance.

In summary, the use of sludge recycle to adjust to changing wastewater conditions can be done within limits if the operator knows how this will affect the SRT value and how it will affect the solids content of the final effluent. In the rotating contactor process, the plant operator has no process adjustments to make, and there is little he can do to inadvertently upset the system. It is the lack of adjustability that gives the process its inherent simplicity and stability. As long as the media continues to rotate, little can be done to upset the process operation seriously.

One area of flexibility for the rotating contactor process which is generally not feasible for the activated sludge process is the ease with which process performance can be evaluated at intermediate points within the system. Samples taken from each RBC stage can be analyzed to determine where there may be problems due to an oxygen limitation, nutrient deficiency, low alkalinity, adverse pH, high mixed liquor solids concentration, poor solids settleability, etc. Interstage sampling is also helpful for determining where nitrification begins and where the majority of the BOD reduction occurs. In an underloaded plant, the operator can determine the point in the system at which he is exactly meeting the treatment requirements. This enables him to estimate closely the number of parallel flow paths of shaft assemblies he may choose to shut down to save operating costs.

MANAGEMENT OF OPERATIONS

Management of rotating contactor process operations can be divided into two categories: 1. wastewater sampling and analysis, and 2. equipment maintenance.

Wastewater Sampling and Analysis

Since there are no process control functions to be performed with the rotating contactor process, analysis is required only to monitor and report daily plant performance. This is done best by analysis of 24-hr composite wastewater samples collected on a daily basis.

Samples of wastewater are usually taken before primary treatment, primary effluent (rotating contactor influent), and final clarifier effluent. Analysis of rotating contactor effluent (final clarifier influent) is also a good practice. This will better separate the performance factors associated with the rotation contactor process and the final clarifier. Analysis of samples, as a minimum, will include BOD_5 and suspended solids. Depending on effluent standards, analysis for NH_3-N, NO_3-N, TKN, and phosphorus may also be included. For the rotating contactor process, personnel skills required to perform the analysis function are reduced, since there is no control or process adjustment function to be performed in conjunction with sample analysis.

Automatic composite sampling equipment will serve to monitor sampling points of key interest to regulatory agencies. Evaluation of plant performance at peak hydraulic and organic loads requires auxiliary automatic sampling equipment. Preferably, this equipment would produce separate samples at hourly intervals. This provides the operator with an hourly profile of wastewater characteristics and the effect of various daily plant operations (sludge dewatering, digester supernatant return, etc.), as well as the ability to develop a flow-proportional composite sample.

Manual sampling can be performed at all points in the system in lieu of automatic sampling devices, whether for regulatory agency monitoring or plant management. Sampling for regulatory agencies will require samples taken at frequent intervals to produce a 24-hr composite. Sampling for plant management purposes can be less frequently performed. In some cases, a single grab sample will suffice. Sampling on an hourly basis during peak hydraulic or organic load periods can also be done manually.

To evaluate the operation of individual stages of a rotating contactor plant, individual mixed liquor samples must be settled to separate mixed liquor solids prior to compositing. This avoids changes in settleability and the interaction of the biological solids and wastewater during the storage period.

Equipment Maintenance

The objectives of an equipment maintenance program are to keep the rotating contactor shafts rotating continuously with minimum down time. Equipment simplicity makes this a relatively easy function to perform. The shafts are rotated by

FIGURE 7-10: Electrical mechanical drive for rotating biological contactor.

electrical — mechanical drive equipment, usually consisting of motor, belt drive, gear reducer, and chain drive, as shown in Figure 7-10. The main shaft is supported by two main bearings. Emphasis is on preventative maintenance. Main bearings must be periodically checked and greased. Chain drive and reducer oil levels are checked periodically, and oil replacement is an occasional requirement. The main bearing housing and drive package will infrequently require preparation and repainting. Chains and belts will be replaced at infrequent intervals. Neither the rotating contactor media nor the main support shaft require maintenance.

Due to the equipment simplicity, personnel do not require specialized skills, and manpower levels are quite low. All routine maintenance functions may be performed by essentially unskilled labor with limited job training. All that is required under preventative maintenance is to keep the chain, reducer, shaft bearings, and motor well lubricated and the chain and belt (if any) drive systems well aligned under the proper amount of tension. Preventive maintenance procedures are summarized in Table 7-1.

TABLE 7-1

Preventive Maintenance Guide

Weekly procedures

1. Check shaft bearings; feel if they are running hot. Listen for unusual noises; this includes any pillow block on the output shaft of the speed reducer.
2. Feel motors to make sure they are running hot.
3. Check the area around the drive train and shaft bearings for oil spills.
4. Check oil levels in speed reducer and chain drive system.

Monthly procedures

5. Lubricate shaft bearings. Consult manufacturer's instructions.

Quarterly procedures

6. Check chain drives for alignment and tightness.
7. Check belt drives (if any) for alignment and tightness.
8. Coat the machined ends of the rotating contactor shaft with grease in case these ends do not have a permanent coating.
9. Adjust shaft bearings; this includes any pillow block on the reducer output.

Semi-annual procedures

10. Change lubricant for chain drive system.

Annual procedures

11. Change oil in speed reducer. Clean magnetic drain plug, if any.
12. Replace the grease in the seals (if any) in the speed reducer. Consult manufacturer's instructions.
13. Grease bearings in the electric motor (if applicable). Consult manufacturer's instructions.

REFERENCE

1. **Wilson, T. and Riddle, M.,** Nitrogen removal: Where do we stand? *Water Wastes Eng.,* 11, 1, 1974.

Chapter 8

OPERATING EXPERIENCE

MUNICIPAL WASTEWATER TREATMENT

0.50-mgd Plant

Pewaukee, Wisconsin is the site of a 0.5-mgd wastewater treatment plant incorporating the rotating contactor process for secondary biological treatment.[1] This is the first full-scale municipal wastewater treatment plant in the U.S. utilizing this treatment process. The treatment plant was constructed and evaluated with EPA demonstration grant funds.

Demonstration Plant

The demonstration plant shown in Figure 8-1 consists of the following principal components:

1. In the foreground is a 50-ft-diameter, combined primary and secondary clarifier. A circular, 36-ft-diameter, 7-ft-deep center section serves as a secondary clarifier, and a 6.5-ft-wide by 7-ft-deep annular ring serves as a primary clarifier. A single rotating bridge with two scraper mechanisms collects settled materials from both the primary and secondary sections. Both sections are designed for a surface overflow rate of 800 gpd/ft^2.

2. In the background to the right is a 50-ft by 60-ft building, housing the rotating contactor equipment. It contains 8 shafts 18 ft long, each with 150 10-ft-diameter polystyrene discs. The discs are fabricated from expanded polystyrene beads, are ½ in. thick, and are spaced on $1\frac{1}{3}$-in. centers. Each disc provides approximately 140 ft^2 of surface area available for biological growth, or 21,000 ft^2/shaft. The shafts are mounted in semicircular-shaped concrete tanks, which closely conform to the shape of the discs and are arranged in two parallel paths of four shafts each. Wastewater flow is perpendicular to the shafts, and each shaft of discs provides an individual stage of biological treatment (see Figure 8-2). Except for tankage contours and interstage weirs, the arrangement is similar to Configuration # 3 in Chapter 4.

The shafts are driven independently by a drive system, consisting of a 1.5-hp motor, helical gear reducer, and chain and sprocket final drive capable of disc-speed variations of 0.75 to 2 rpm. The building protects the discs from damage due to wind, precipitation, or vandalism, and protects the biological growth from freezing temperatures. The construction and operation of this installation are identical to a typical European plant.

3. In the background and to the left in Figure 8-1 is a covered, aerobic digester with a floating 20-hp surface aerator. It is designed for a 20-day solids-aeration time for combined primary and secondary sludge.

4. Sludge drying beds (not shown).

5. Chlorination facilities (not shown).

Average design flow for the treatment plant is 470,000 gpd. The primary clarifier is designed for a BOD removal of 30%, while the rotating contactor process portion is designed for 86%, for an overall BOD removal of 90%.

The treatment plant operates as follows: Raw wastewater passes through a Parshall flume and comminutor to a wet well. Wastewater is pumped from the wet well to the primary portion of the combined primary and secondary clarifier. Primary effluent flows by gravity to the rotating contactor building, where it is divided between the two parallel paths of treatment. Wastewater is distributed along the length of the first shaft of discs by a V-notch weir. Mixed liquor in each stage of treatment flows over a flat-edge weir to each subsequent stage. Total head loss through the 4 stages of treatment is approximately 4 in. Effluents from the two parallel paths of treatment are combined and flow by gravity to the secondary clarifier. Effluent from the secondary clarifier is chlorinated prior to discharge to the Pewaukee River. Secondary sludge is drawn from the clarifier by an automated valve and flows back to the wet well of the plant. Raw sewage pumps return secondary sludge to the primary clarifier where it settles and combines with the primary sludge, and together they are pumped to the aerobic digester. Digester supernatant is drawn off with a telescoping valve and flows to the wet well of the plant. Digested sludge is dewatered on drying beds prior to disposal by land fill.

Test Program

The Pewaukee treatment plant began operation in December 1971, and underwent a full year of evaluation as part of the demonstration program.

FIGURE 8-1. RBC demonstration plant at Pewaukee, Wisconsin. (From Antonie, R. L., Kluge, D., and Mielke, J., *J. Water Pollut. Control Fed.*, 46(3), 498, 1974. With permission.)

FIGURE 8-2. Interior of RBC building at Pewaukee, Wisconsin. (From Antonie, R. L., Kluge, D., and Mielke, J., *J. Water Pollut. Control Fed.*, 46(3), 498, 1974. With permission.)

Variables investigated included rotational disc velocity, hydraulic loading, and exposure to different climatic conditions. The data reported here cover the period of testing from December 1971 through August 1972.

Unless otherwise indicated, all samples taken are refrigerated composite samples. Primary and secondary clarifier effluent samples are 24-hr composites. Samples from the individual stages of discs are composites of three grab samples taken in the morning, noon, and afternoon of each day of sampling. All analyses were performed within 8 hr of the end of the sampling period using the procedures outlined in *Standard Methods for examination of Water and Wastewater, 13th Edition.*

The data presented in the following correlations are from individual daily samples. Unless otherwise indicated, no averaging of test results was performed. Percent reductions of the various wastewater components are stated on the basis of primary clarifier effluent and do not include any

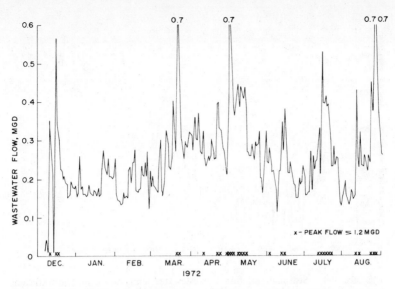

FIGURE 8-3. Daily wastewater flow variations in demonstration tests at Pewaukee, Wisconsin. (From Antonie, R. L., Kluge, D., and Mielke, J., *J. Water Pollut. Control Fed.*, 46(3), 498, 1974. With permission.)

additional removal normally attributed to primary treatment.

Wastewater Characteristics

Average daily wastewater flow to the demonstration plant for the first 9 months of operation is shown in Figure 8-3. After the plant was put into full operation by mid-December, the wastewater flow remained relatively uniform for the winter months. However, from mid-March through August, the wastewater flows were highly variable due to sewer infiltration from spring thawing and rainfall. The ratio of daily peak to average flows for the 9-month period varied from 2.5 to 4. The notations along the abscissa in Figure 8-3 indicate days on which the peak flow equaled or exceeded 1.2 mgd, which is capacity for the flow-measurement equipment.

Because the lift station at the treatment plant had to be designed to handle peak wet weather flows from sewer infiltration, the pumps were somewhat larger than would usually be used for a plant of this size. At normal flows the oversized pumps produced a pulsed input to the treatment plant, because they would pump at a very high rate until the wet well was pumped down and then shut off until the wet well filled. This pulsed operation introduced flow variations much greater than shown in Figure 8-3 and is likely to have affected both secondary treatment and clarification efficiency.

The temperature of the primary clarifier effluent is shown in Figure 8-4. Wastewater temperature measurements began in mid-December and showed levels below 55°F. Cold weather reduced the wastewater temperature into the mid-forties, where it remained throughout the winter. Sewer infiltration from spring thawing and rainfall kept the wastewater temperatures in the mid-forties until mid-April. From there, they increased gradually, but did not remain consistently above 55°F until the end of May. The influence of wastewater temperature on process efficiency is discussed in a later section.

The BOD concentration of the wastewater at Pewaukee, Wisconsin, was relatively low. Figure 8-5 shows that during the winter months, the primary clarifier effluent BOD concentration ranged from 120 to 160 mg/l. However, sewer infiltration during the spring and summer months resulted in a decrease in the BOD concentration to an average value of approximately 100 mg/l. From the first day of rotating contactor operation (December 6), the secondary clarifier effluent BOD concentration decreased rapidly to a secondary treatment quality effluent, indicating that the process started up and approached steady state operation within 1 to 2 weeks. During the first 5 weeks of operation, the valve which allows secondary sludge to be removed from the final clarifier was not operable. On January 6, this valve was made operable, and 5 weeks of accumulated

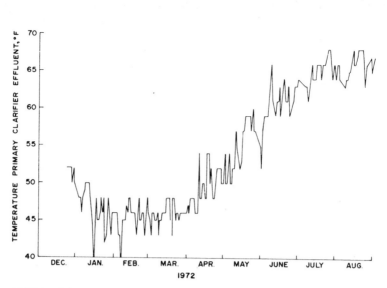

FIGURE 8-4.   Seasonal variation in wastewater temperature in demonstration tests at Pewaukee, Wisconsin. (From Antonie, R. L., Kluge, D., and Mielke, J., *J. Water Pollut. Fed.,* 46(3), 498, 1974. With permission.)

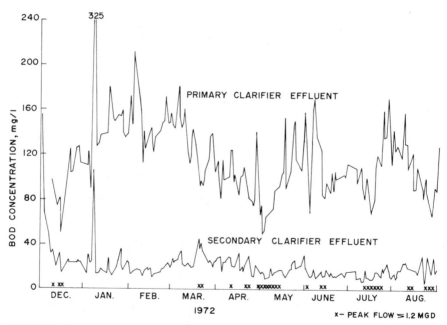

FIGURE 8-5.   BOD data for secondary treatment tests at Pewaukee, Wisconsin. (From Antonie, R. L., Kluge, D., and Mielke, J., *J. Water Pollut. Control Fed.,* 46(3), 498, 1974. With permission.)

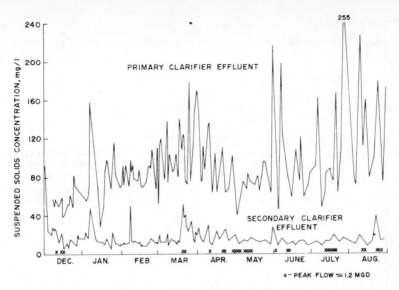

FIGURE 8-6. Suspended solids data for secondary treatment tests at Pewaukee, Wisconsin. (From Antonie, R. L., Kluge, D., and Mielke, J., *J. Water Pollut. Control Fed.,* 46(3), 498, 1974. With permission.)

secondary sludge was removed from the secondary clarifier to the wet well of the plant, which resulted in the 1-day peak in the secondary clarifier effluent BOD concentration. From then on, the secondary clarifier effluent had a relatively uniform BOD concentration, despite the cold wastewater temperatures during the winter and the highly variable influent BOD concentrations during the summer. The days indicated with peak wastewater flows of 1.2 mgd or more had peak secondary clarifier surface overflow rates of 2,000 gpd/ft² or more. This apparently had only a slight effect on secondary clarifier effluent quality.

Figure 8-6 shows suspended solid concentrations in the primary and secondary clarifier effluents. From the first day of operation, suspended solids concentrations in the secondary clarifier effluent also rapidly approached secondary treatment quality. Except for one day in February and a few days in March, the secondary clarifier effluent remained relatively uniform despite low wastewater temperatures in the winter and spring and highly variable influent concentrations in the spring and summer. The peak flows of 1.2 mgd or greater also did not result in any significant carry-over of suspended solids from the secondary clarifier.

Process Variables

To obtain operating data from the demonstration plant over a wide range of hydraulic loading, the wastewater flow was split unevenly between the two parallel paths of treatment during the summer months. The ratio of the flow split ranged between 2 and 4 to 1 with the northern half of the plant receiving the lower flow. To evaluate process performance, samples of effluent were taken from each parallel path of treatment, allowed to settle for 30 min, and composited for analysis. BOD removal as a function of hydraulic loading is shown in Figure 8-7. Because the primary effluent BOD concentration varied considerably during the testing, these data in Figure 8-7 have been divided into two concentration ranges. The higher concentration range is a more typical primary effluent. In this range of concentration, the process achieved 85% BOD removal at a hydraulic loading of approximately 0.23 mgd or a disc surface area loading of 2.8 gpd/ft².

As wastewater flow was decreased, BOD removal increased until, at a flow of 0.04 mgd, 95% BOD removal was obtained. Effluent BOD concentrations for 85% removal varied from 15 to 25 mg/l BOD, and at 95% removal, from 5 to 10 mg/l BOD.

Data from the lower concentration range in Figure 8-7 are widely scattered and show a decrease in the treatability of the wastewater. This is due partly to the limits on effluent BOD concentration obtainable with biological treatment, and also to the earlier beginning of nitrification which masks the carbonaceous BOD removal.

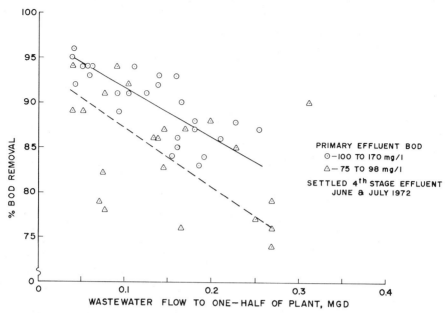

FIGURE 8-7. Correlations of BOD removal for Pewaukee, Wisconsin demonstration tests. (From Antonie, R. L., Kluge, D., and Mielke, J., *J. Water Pollut. Control Fed.*, 46(3), 498, 1974. With permission.)

The scatter in the data is also due to the difficulty in achieving valid comparisons of influent and effluent concentrations when rainfall and sewer infiltration cause wide fluctuations in consecutive daily influent BOD concentrations.

If the two lines of correlation in Figure 8-7 are compared to equal operating conditions in Figure 4-1, it may be found that they do not indicate equal performance levels. This is due mainly to a low volume to surface area ratio, but also in part to the pulsed flow operation. Because the tankage closely conforms to the shape of the discs, there is a relatively small column of wastewater in the process, about 0.85 gal/ft² of disc surface area. The data in Figure 4-1 are from operation with larger tankage with a 0.12 gal/ft² volume to surface area ratio. The 50% longer wastewater retention time yields a significant increase in treatment capacity (see the discussion of volume to surface ratio in Chapter 3).

The effect of wastewater flow on nitrification is shown in Figure 8-8. Although this plant was not designed for ammonia nitrogen removal, nitrification occurred when it was operated at flows below design flow. For the higher range of primary effluent BOD concentration, nitrification began at a flow of approximately 0.15 to 0.2 mgd. As the flow was decreased, the nitrification proceeded

rapidly until at the flow of approximately 0.04 mgd, nitrification was complete. For the lower BOD concentration range, nitrification began at a flow of about 0.25 mgd; as the flow was decreased, nitrification continued until at a flow of approximately 0.05 mgd, nitrification was complete. Previous testing has shown that nitrification begins in the process when the wastewater BOD concentration approaches 30 mg/l. At this concentration, nitrifying organisms can compete with the more rapidly growing carbon-oxidizing organisms and can establish themselves in the process. Nitrification then proceeds rapidly until, at a BOD concentration of approximately 10 mg/l, the nitrification is complete. This occurs independently from the initial BOD concentration and is the reason why nitrification began at a higher flow for the lower concentration range in Figure 8-8.

Comparing results from Figure 8-8 and Figure 4-7 indicates lower performance by the Pewaukee plant. Again this is due to the pulsed flow and low volume to surface area ratio, however, in this case, the pulsed flow is much more important because of possible displacement of nitrifying bacteria from the latter stages.

Ammonia nitrogen concentration in the raw wastewater at Pewaukee averaged approximately 15 mg/l. The effluent ammonia nitrogen concen-

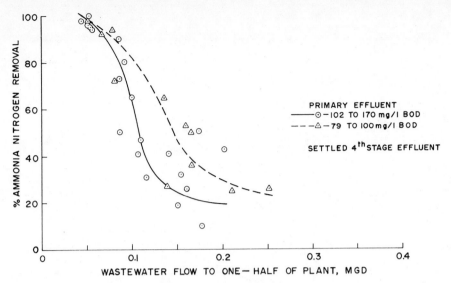

FIGURE 8-8. Nitrification correlation for Pewaukee, Wisconsin demonstration tests. (From Antonie, R. L., Kluge, D., and Mielke, J., *J. Water Pollut. Control Fed.*, 46(3), 498, 1974. With permission.)

trations obtained are shown in Figure 8-9 as a function of hydraulic loading. For the higher BOD concentration range, effluents of less than 1 mg/l ammonia nitrogen were produced at a wastewater flow of approximately 0.05 mgd. For the lower BOD concentration range, effluent ammonia nitrogen concentrations of less than 1 mg/l were produced at a wastewater flow of approximately 0.08 mgd.

The preceding data correlations were all based on wastewater temperatures of 55°F or above. Previously it was shown that the wastewater temperature at Pewaukee during the winter and spring months is considerably below this level. Figure 8-10 shows the effect of wastewater temperature on BOD removal by the treatment plant for all seasons of operation. The data have been divided into four different flow ranges to show the relationship of hydraulic loading and wastewater temperatures with BOD removal efficiency. BOD removals for the lowest wastewater flow are shown by the uppermost line in Figure 8-10. For normal wastewater temperatures of 55°F and above, approximately 91% removal was achieved. As the wastewater temperature decreased to approximately 45°F, the BOD removal decreased to approximately 88%. The same pattern is shown for the other flow ranges. As the wastewater flow increased, the inhibition caused by low wastewater temperatures also increased. For the highest flow range, BOD removals of approximately 85% at

normal wastewater temperatures decreased to about 78% at 45°F. Figure 8-10 indicates that it is easy to compensate for low wastewater temperatures by designing a plant for an appropriately lower hydraulic loading. For example, if it is desired to obtain 85% BOD removal at a wastewater temperature of 45°F, it would be necessary to design the plant for approximately 87 to 88% BOD removal at normal wastewater temperatures. When designing plants for high degrees of BOD removal, the longer retention time of the wastewater (and longer sludge age) decreases the inhibitory effect of the low temperatures; and a somewhat smaller adjustment in design hydraulic loading is sufficient to maintain the treatment level.

Except for a few tests during February and March, the operating data presented in this report were collected at a rotational disc velocity of 2 rpm or a peripheral velocity of approximately 60 ft/min. Previous testing has shown the peripheral velocity of 60 ft/min to optimize the removal of both BOD and ammonia nitrogen. In these studies, all stages of discs rotated at the same velocity. It seems possible that the lower aeration requirements in the later stages of treatment would allow a lower rotational disc velocity.

Process Effluent Characteristics

At hydraulic loadings where the rotating contactor process is achieving ammonia nitrogen

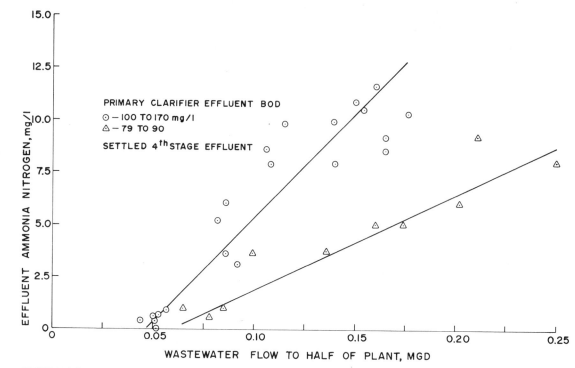

FIGURE 8-9. Effluent ammonia correlation for Pewaukee, Wisconsin demonstration tests. (From Antonie, R. L., Kluge, D., and Mielke, J., *J. Water Pollut. Control Fed.*, 46(3), 498, 1974. With permission.)

removal, the effluent from the process is heavily seeded with nitrifying organisms. Because of this, there is a significant amount of nitrogenous oxygen demand exerted during the 5-day incubation period of the BOD test. To determine the proportions of carbonaceous and nitrogenous oxygen demand in the rotating contactor effluent, BOD analyses were also conducted where 0.5 mg/l of allylthiourea was added to the BOD dilution water to suppress nitrification during the incubation period. The results of these tests are shown in Figure 8-11. In the BOD_5 concentration range of 15 to 30 mg/l, anywhere from 20 to 50% of the oxygen demand is nitrogenous, and 50 to 80% is carbonaceous. As the BOD_5 concentration is reduced into the range of 5 to 10 mg/l, most of the ammonia nitrogen is removed, and little nitrification occurs during the BOD test. In this range, BOD_5 removals of 85 to 90% will correspond to carbonaceous BOD removals as high as 90 and 95%, respectively.

Ammonia nitrogen removal by the process consists primarily of oxidation to nitrite and nitrate nitrogen. Figure 8-12 shows the conversion of ammonia nitrogen to nitrate as a function of the amount of ammonia nitrogen removed. Samples taken from the final stage of treatment

indicate that about 90% of the ammonia nitrogen is oxidized to the nitrate form over the entire range of ammonia nitrogen removals. However, samples taken of the final clarifier effluent indicate that anywhere from 30 to 50% denitrification occurs as the fourth-stage effluent passes through the secondary clarifier. This occured despite the presence of significant amount of dissolved oxygen in the final clarifier effluent. Apparently the denitrification occurred at the bottom of the clarifier in the area of the settled sludge.

Figure 8-13 shows the relationship between effluent ammonia nitrogen and effluent Kjeldahl nitrogen concentrations from the process. It indicates that Kjeldahl nitrogen (including ammonia nitrogen) is removed in direct proportion with ammonia nitrogen removal. The relationship holds over the entire range of effluent ammonia nitrogen concentrations, and at essentially complete nitrification, approximately 2 mg/l of organic nitrogen remain in the effluent. These are accounted for partly in the remaining suspended solids, and partly in residual colloidal and soluble organic matter which appears to be very resistant to biological degradation.

The primary mechanisms of organic nitrogen

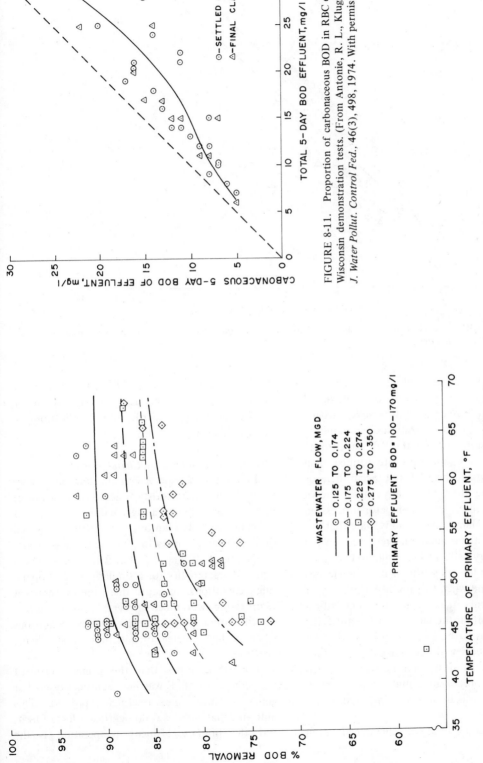

FIGURE 8-11. Proportion of carbonaceous BOD in RBC effluent in Pewaukee, Wisconsin demonstration tests. (From Antonie, R. L., Kluge, D., and Mielke, J., *J. Water Pollut. Control Fed.*, 46(3), 498, 1974. With permission.)

FIGURE 8-10. Effect of wastewater temperature on BOD removal in Pewaukee, Wisconsin demonstration tests. (From Antonie, R. L., Kluge, D., and Mielke, J., *J. Water Pollut. Control Fed.*, 46(3), 498, 1974. With permission.)

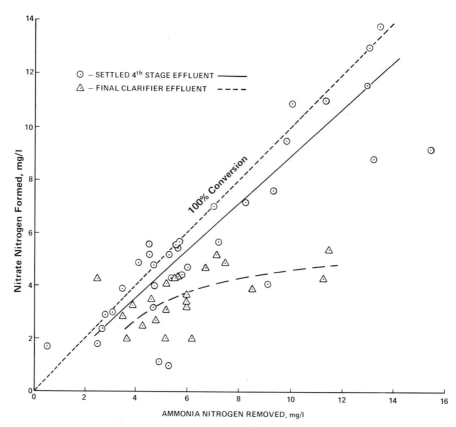

FIGURE 8-12   Conversion of ammonia nitrogen to nitrate nitrogen in Pewaukee, Wisconsin demonstration tests.

removal are flocculation and settling and cell synthesis. It is often suggested that biological treatment results in hydrolysis of organic nitrogen to ammonia followed by oxidation of ammonia to nitrate. If this were true for the RBC, a nitrogen balance would show more nitrate formed than ammonia removed. This did not occur here as shown in Figure 8-12 or in any other known investigation of the rotating contactor process on domestic waste. Many times there is somewhat less nitrate formed than ammonia removed. When treating a primary effluent this is due to consumption of ammonia for cell synthesis in carbonaceous BOD removal, and when treating either primary or secondary effluents it is due to some denitrification occurring within the biomass of the nitrifying stages. When this is the case, hydrolysis followed by oxidation occurs without being detected in a nitrogen balance. In most domestic waste applications, however, only 1 to 2 mg/l of the organic nitrogen present will be involved in a hydrolysis-oxidation reaction.

Figures 8-12 and 8-13 are of value in estimating the total nitrogen content of the effluent from the process and should be used to establish design requirements for denitrification applications as discussed in Chapter 4.

Mixed Liquor and Sludge Characteristics

Tables 8-1, 2, and 3 contain data on average mixed liquor characteristics measured under three different sets of operating conditions. Table 8-1 indicates the mixed liquor characteristics measured during winter operation at low wastewater temperatures. Table 8-2 shows the same mixed liquor characteristics for a similar period of operating conditions, except that the wastewater temperature was much higher. Table 8-3 shows mixed liquor characteristics under conditions where a high degree of nitrification was being achieved.

Table 8-1 shows that the primary effluent temperature of 46°F was not reduced any further going through the treatment process. This indicates that there was no significant loss of heat, and the primary effluent temperature is the

TABLE 8-1

Pewaukee, Wisconsin Demonstration Program
Mixed Liquor Characteristics

January 13 to April 9, 1972
Wastewater Flow = 0.15 to 0.7 mgd

	Temperature		pH		D. O.		BOD		SS	
	°F	N	pH	N	mg/l	N	mg/l	N	mg/l	N
Primary effluent	46	87	7.7	87	4.3	58	136	67	93	59
Stage 1	–	–	–	–	3.1	60	86	12	82	8
2	–	–	–	–	3.5	60	40	12	84	8
3	–	–	–	–	4.7	60	24	12	105	8
4	47	87	7.7	87	6.7	116	23	82	72	30
Final effluent	–	–	7.6	87	6.2	58	21	64	16	59

N – Number of determinations.

From Antonie, R. L., Kluge, D., and Mielke, J., *J. Water Pollut. Control Fed.,* 46(3), 498, 1974. With permission.

TABLE 8-2

Pewaukee, Wisconsin Demonstration Program
Mixed Liquor Characteristics

May 16 to August 31, 1972 – Southern half of the plant
Wastewater Flow = 0.1 to 0.6 mgd

	Temperature		pH		D. O.		BOD		SS	
	°F	N	pH	N	mg/l	N	mg/l	N	mg/l	N
Primary effluent	63	74	7.6	89	1.1	71	114	82	105	42
Stage 1	–	–	–	–	1.0	43	63	14	89	13
2	–	–	–	–	1.2	43	40	14	72	13
3	–	–	–	–	1.9	43	29	14	79	13
4	63	74	7.7	89	2.1	71	19	56	80	28
Final effluent	–	–	7.6	89			16	68	14	28

N – Number of determinations.

From Antonie, R. L., Kluge, D., and Mielke, J., *J. Water Pollut. Control Fed.,* 46(3), 498, 1974. With permission.

temperature which controls process efficiency. The pH of the wastewater going through the plant did not change significantly. Because of the low wastewater temperature, the primary effluent had a relatively high D.O. concentration. This decreased slightly in the first stage of treatment, because of the increased biological activity. As the wastewater continued through the process, the dissolved oxygen concentration gradually increased to a relatively high level due to high solubility and relatively low biological activity at the low wastewater temperatures. The dissolved oxygen in the final clarifier effluent was only slightly lower than the effluent from the fourth stage of treatment. The stage-by-stage settled BOD samples seem to indicate that relatively little treatment occurred in the fourth stage of discs. This, however, was not the case, because the initial stages of discs removed only carbonaceous BOD, while the final two stages performed a moderate degree of nitrification. The nitrification which occurs during the 5-day BOD test also masks the

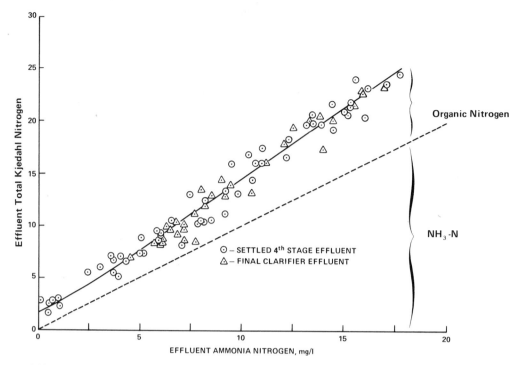

FIGURE 8-13. Distribution of ammonia and Kjeldahl nitrogen in RBC effluent. (From Antonie, R. L., Kluge, D., and Mielke, J., *J. Water Pollut. Control Fed.*, 46(3), 498, 1974. With permission.)

TABLE 8-3

**Pewaukee, Wisconsin Demonstration Program
Mixed Liquor Characteristics**

June 1 to August 10, 1972 – Northern half of the plant
Wastewater flow = 0.04 to 0.07 mgd

	Temperature		D. O.		NH_3-N		pH		BOD	
	°F	N	mg/l	N	mg/l	N	pH	N	mg/l	N
Primary effluent	63	26	1.3	26	14.1	13	7.6	25	118	23
Stage 1	–	–	2.9	14	11.3	6	–	–	20	4
2	–	–	3.4	14	3.2	6	–	–	18	4
3	–	–	5.2	14	0.6	6	–	–	12	4
4	63	26	6.2	26	0.4	13	7.7	25	11	20

N – Number of determinations.

From Antonie, R. L., Kluge, D., and Mielke, J., *J. Water Pollut. Control Fed.*, 46(3), 498, 1974. With permission.

TABLE 8-4

Pewaukee, Wisconsin Demonstration Program
Sludge Solids Content

Secondary clarifier draw-off schedule		Secondary sludge		Combined primary and secondary sludge	
Frequency, times per day	Duration, Min	Percent solids	Number of samples	Percent solids	Number of samples
1	2	4.2	6	4.3	5
2	2	2.0	3	5.0	5
2	3	5.0	2	6.5	2
3	4	4.7	2	5.2	2
4	4	1.5	13	4.8	13
8	2	3.2	3	5.1	3
12	2	2.2	4	4.2	4

From Antonie, R. L., Kluge, D., and Mielke, J., *J. Water Pollut. Control Fed.,* 46(3), 498, 1974. With permission.

carbonaceous BOD removal, which does continue in the final stages of treatment.

Mixed liquor suspended solids concentrations in all four stages of treatment were in the range of 100 mg/l or less. Suspended solids of 72 mg/l left the final stage of treatment and flowed to the secondary clarifier. The effluent from the secondary clarifier contained 16 mg/l suspended solids, indicating that only 56 mg/l of sludge were generated by the secondary treatment process. The low mixed liquor solids concentration generated by the process allows flexibility in secondary clarifier design, because discrete particle settling controls solids separation efficiency and hindered settling and sludge compression are not considerations.

Table 8-2 shows mixed liquor characteristics for the southern half of the demonstration plant under similar hydraulic loadings during the summer. The temperature and pH of the primary effluent underwent no significant change going through the treatment process. Because of the lower oxygen solubility and higher biological activity, the dissolved oxygen concentration of the primary effluent was considerably lower. The D.O. concentration increased gradually in the process to a level just over 2 mg/l. Final clarifier effluent D.O. is not shown, since it was much higher than the fourth-stage effluent. This occurred, because the northern half of the plant was being operated at a significantly lower hydraulic loading and produced a very high effluent D.O. concentration. Percentage removals of BOD and suspended solids

were higher due to the warmer wastewater. Stage-by-stage reductions in BOD suspended solids were similar to those for cold weather conditions.

Table 8-3 shows mixed liquor characteristics for the northern half of the plant when operated at a low wastewater flow. Because of the longer wastewater retention time, the relatively low primary effluent dissolved oxygen concentration was increased rapidly through all four stages of treatment to a level in excess of 6 mg/l. Ammonia nitrogen removal began in the first stage to a moderate degree and proceeded very rapidly through Stages 2 and 3. Only a moderate additional removal occurred in the fourth stage. The high alkalinity of the Pewaukee wastewater prevented any significant change in pH going through the process, even though a high degree of nitrification was occurring. BOD tests on the individual stages indicated that the majority of the carbonaceous matter was removed in the first stage of treatment, and relatively little was removed in the following three stages, where the attached culture was predominated by nitrifying organisms.

The sludge collection mechanism in the secondary clarifier thickened settled sludge to a relatively high solids content. Table 8-4 shows the solids content of the secondary and combined primary and secondary sludges as they are affected by the secondary sludge removal procedure. Secondary sludge solids contents of 4 to 5% were produced when the sludge was removed infrequently and for short durations. As the frequency and duration of sludge removal increased, the secondary sludge solids were diluted into the range of 1 to 2% solids. This occurred because significant amounts of clarifier overflow were included in the sludge drawoff. If it is desired to produce a secondary sludge of high solids content, it is recommended that sludge be withdrawn in a more controlled fashion. This can be done best with a plunger or piston-type sludge pump operated automatically by a timer.

When the secondary sludge was returned to the primary clarifier, and settled and thickened along with the primary sludge, a combined sludge solids content of 4.2 to 6.5% was produced. The solids content of the combined sludges is apparently unaffected by the procedure of removing the sludge from the secondary clarifier. The high sludge solids content obtainable with the rotating contactor process indicates that for many applications sludge thickening prior to treatment and disposal is not necessary. This will result in considerable capital and operating cost savings for a treatment plant.

Operation and Maintenance

Because the rotating contactor process is inherently stable under conditions of fluctuating hydraulic and organic loads, and it does not require recycle of sludge or effluent for proper operation, there need not be special operating flexibility and sophisticated instrumentation designed for the treatment plant. The only instrumentation provided in the Pewaukee demonstration plant are flow measurement and recording equipment, eight starters for the drive motors, and a timer to operate the secondary sludge drawoff valve.

The use of simple mechanical components for the process hardware results in a very low requirement for mechanical maintenance. This allows a relatively low level of operator skill to be used for the treatment plant. For small- and medium-sized communities, this is very important, because they often cannot afford nor obtain skilled treatment plant operators. For larger communities, this allows treatment plant operators to spend more time on other aspects of wastewater treatment plant operation, which are prone to be more troublesome. Use of separate drive systems for each shaft of media allows an individual shaft to be shut down for maintenance without affecting operation of the balance of the treatment plant. Wastewater flow continues through the plant

without receiving treatment by the nonrotating shaft and only a moderate loss of treatment efficiency is experienced.

Tertiary Filtration

To determine how amenable RBC effluent is to filtration as a means of tertiary treatment, a pilot-scale, dual media pressure filter was tested at the treatment plant at Pewaukee, Wisconsin. The pilot filter consisted of a 14-in. diameter by 60-in.-high tank containing different proportions of 0.5 mm of sand and anthracite in a total bed depth of 30 in. The bottom of the tank contained 8.5 in. of gravel as an underdrain, leaving 22.5 in. at the top for bed expansion during backwash. Backwashing was performed manually at convenient intervals at approximately 15 gpm/ft^2 for approximately 10 min and usually represented about 5% of total flow.

Effluent from the secondary clarifier following the RBC process was pumped through the filter at rates of 2.5 to 4.5 gpm/ft^2 and resulted in filter runs of 8 to 24 hr. Composite samples were taken of the filter influent and effluent. These samples were again filtered in the laboratory through media of various porosities and further BOD analyses were performed to determine the particle size distribution of the remaining BOD. The results of the dual media filtration tests for BOD and suspended solids removal are presented in Figure 8-14.

The data from the tests described above show that RBC effluent of 25 mg/l BOD or less contains 70 to 80% insoluble BOD as measured by a 0.45-μ filter and that about 50% of the effluent BOD can be removed by a dual media pressure filter operated at conventional loadings. Similar removals would be anticipated with other types of filtration equipment, such as gravity sand filters and microscreens. These tests were performed without adding chemicals to aid flocculation and filtration.

1.0-mgd Plant

The first rotating contactor plant to be built in the U.S. with EPA construction grant funds is at Gladstone, Michigan. It has been in operation since March 1974.[2] This application of the RBC process is for upgrading an existing primary plant to secondary treatment at wastewater temperatures as low as 46°F.

The existing plant at Gladstone consisted of

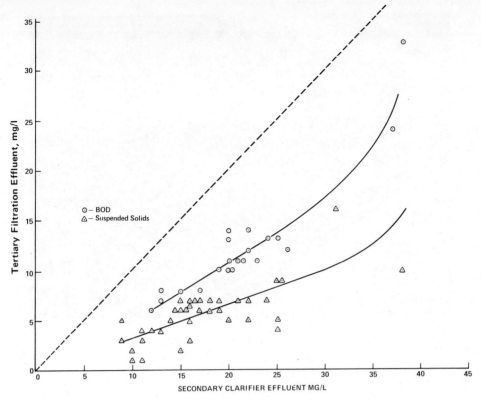

FIGURE 8-14. Filtration of RBC effluent for Pewaukee, Wisconsin demonstration tests.

two rectangular primary clarifiers inside a building. The secondary treatment addition consists of two parallel paths of three RBC media assemblies each and an extension of the building to enclose them. Wastewater flow is parallel to the shafts and each shaft is divided into two stages providing a total of six stages in each flow path (this is similar to Configuration #2 in Chapter 7).

Figure 8-15 is a view of the 6 rotating contactor shafts which contain a total of 516,000 ft² of surface area.

Other plant additions include two rectangular secondary clarifiers designed at 600 gpd/ft², expanded anaerobic digestion facility which includes two 16,000 ft³ heated digesters operated in series, and expanded chlorination and sludge drying bed facilities.

Raw Wastewater Characteristics

Raw wastewater data in Table 8-5 collected between March and December 1974 show that the wastewater flow was fairly consistent from month to month and averaged 0.72 mgd. During the winter of 1975 it decreased to about 0.5 mgd but increased in spring to about 1.0 mgd. Daily

peak-to-average flow ratio varied between 2.0 and 1.5 to 1.0. When the plant started operation in March, the wastewater temperature was quite low (46°F). The temperature gradually increased to the low-to-middle 60s by mid-summer, and then gradually decreased to the mid 40s the following winter. BOD and suspended solids concentrations remained consistent throughout the operating period. The average influent BOD and suspended solids concentrations were 168 mg/l and 133 mg/l, respectively.

Secondary Treatment System

Data in Table 8-6 show that BOD removal by the secondary treatment system gradually improved from start-up in March through August 1974, and from September through December the effluents were consistently less than 10 mg/l BOD. Normally a rotating contactor treatment plant will reach steady state operation within a few weeks after start-up. However, because the influent wastewater temperature was quite cold at start-up, it took several months to reach steady state operation.

During the winter months of 1975 the effluent

FIGURE 8-15   Interior of RBC plant at Gladstone, Michigan.

TABLE 8-5

Raw Wastewater Characteristics at Plant Located in Gladstone, Michigan

Month (1974)	Average flow mgd	gpd ft²	Temperature °F	BOD, mg/l	SS, mg/l	VSS, mg/l	Total P, mg/l
March	0.623	1.21	46	219	126	106	–
April	0.762	1.48	47	166	103	86	–
May	0.731	1.42	53	134	167	120	7.8
June	0.780	1.51	57	143	111	85	6.7
July	0.873	1.69	62	166	123	96	7.3
August	0.774	1.50	65	180	125	100	7.6
September	0.802	1.55	62	155	124	92	6.8
October	0.597	1.16	58	160	155	111	7.1
November	0.676	1.31	55	157	113	87	6.9
December	0.563	1.09	51	175	136	108	8.4
(1975)							
January	0.491	.95	49	203	208	142	9.0
Test period	0.486	1.88	47	184	119	–	–
February	0.510	.99	47	190	152	122	9.1
March	0.619	1.20	47	181	161	119	8.7
April	0.986	1.91	47	153	120	90	7.1
May	0.903	1.75	52	129	101	78	7.4
June	1.044	2.02	59	154	113	86	6.7
July	0.843	1.63	64	214	146	109	8.7
August	0.782	1.52	68	184	168	130	9.3
September	0.888	1.72	66	168	126	75	7.5

TABLE 8-6

Secondary Treatment System Operation at Gladestone, Michigan

Month (1974)	BOD, mg/l Influent	BOD, mg/l Effluent	Solids, mg/l Influent SS	Influent VSS	Effluent SS	Effluent VSS	Effluent D. O. mg/l	Total P, mg/l Influent	Total P, mg/l Effluent	lb Alum*/lb P — lb P	Chlorine dosage, mg/l
March	133	60	62	48	20	–	–	–	–	–	6.16
April	101	42	63	51	21	13	–	–	–	–	5.66
May	78	24	59	47	18	14	–	5.8	4.3	–	6.77
June	99	17	52	37	17	10	–	5.2	3.2	–	7.38
July	105	19	52	38	16	11	5.6	5.1	3.0	–	6.87
August	102	12	52	39	17	13	6.3	5.7	3.4	–	7.44
September	96	8	35	26	20	10	6.5	3.3	1.7	17.3	6.13
October	78	7	60	44	12	7	6.5	5.2	2.2	19.6	4.82
November	103	3	45	37	16	8	7.1	5.2	1.7	21.2	3.19
December	122	5	66	45	16	7	6.9	6.7	2.7	18.2	2.47
(1975)											
January	132	10	94	59	9	6	6.2	7.4	5.7	–	3.66
Test period	137	19	65	49	13	8	5.9	–	–	–	–
February	135	13	75	48	9	5	7.6	6.6	3.3	14.5	3.85
March	137	13	108	65	15	9	6.9	6.9	1.7	12.9	2.9
April	111	8	70	48	19	9	7.9	5.6	1.5	16.9	1.8
May	115	4	75	46	14	6	7.0	6.6	1.4	12.7	1.59
June	122	6	72	51	13	6	6.7	5.6	1.1	15.2	1.95
July	154	9	76	51	12	5	6.2	7.4	1.4	11.1	1.83
August	124	7	97	63	12	7	6.4	8.4	1.6	10.6	2.76
September	126	6	76	49	14	6	6.3	6.5	1.4	12.6	2.03

*Based on P removed only in secondary treatment.

increased to 10 to 13 mg/l BOD as the temperature decreased to 47°F. A 2-week test period was conducted between January and February 1975, where one of the two parallel paths of rotating contactors was shut down along with one of the two final clarifiers to simulate operation at full design flow of 1.0 mgd and low wastewater temperature. The treatment requirement of 80% BOD removal or 30 mg/l effluent BOD was easily met. Effluent suspended solids remained consistent throughout this operating period at an average of about 16 mg/l.

The hydraulic profile of the treatment plant required that the rotating contactor process effluent be pumped to the final clarifiers, rather than flow by gravity as is normally practiced. Solids settleability tests conducted before and after pumping indicated that the solids settleability was somewhat deteriorated by the shearing action of the centrifugal pump. Normally the effluent suspended solids will be approximately the same value as the effluent BOD.

Beginning in September 1974, the effluent BOD values were usually lower than effluent suspended solids. However, at the same time, the volatile solids content of the effluent solids changed significantly. From March through August 1974, the effluent suspended solids were approximately 70% volatile. Beginning in September, this decreased to an average of about 50%. The increase in the inert solids content of the effluent can be attributed to the presence of nonsettleable chemical solids resulting from the addition of alum for phosphorus removal. Effluent D.O. concentrations were consistently about 5 mg/l, even during the warmer summer months.

Beginning in September 1974, alum was added for phosphorus removal. Prior to alum addition, effluent phosphorus concentrations of 3 to 4 mg/l were produced. After alum addition, phosphorus levels in the effluent were generally less than 2 mg/l and as low as 1.1 mg/l. Initally, alum was added to the primary clarifier and about 20 lb

TABLE 8-7

Sludge Production Data for RBC Installation at Gladstone, Michigan Combined Primary and Secondary Sludge

Month (1974)	Daily volume gal × 100	Percent solids	Percent volatile	lb/lb BOD removed	pH
March	12	3.3	74	0.40	—
April	23.2	4.1	76	1.00	—
May	30	3.8	77	1.42	—
June	—	3.5	75	—	—
July	23	3.6	73	0.646	6.1
August	22	3.7	75	0.626	6.2
September	30.3	3.8	70	0.977	6.3
October	35	3.3	72	1.26	6.4
November	42.9	2.2	72	0.907	6.4
December	42.9	2.3	75	1.03	6.5
(1975)					
January	29.4	2.0	79	0.620	6.3
February	24.8	3.9	75	1.07	6.5
March	15.1	4.7	72	0.682	6.4
April	21.0	4.3	70	0.632	6.3
May	20.4	4.0	68	0.723	6.2
June	22.1	4.5	68	0.644	7.0
July	28.3	4.1	64	0.671	6.1
August	31.3	4.4	61	0.995	6.2
September	32	3.9	63	0.867	6.2

alum/lb phosphorus removed were used by the secondary treatment portion of the plant. Based on the overall plant, 12 lb alum/lb phosphorus removed were used. In 1975, alum addition was switched to the effluent from the RBC system. This reduced alum requirements to 12 lb/lb phosphorus removed by secondary treatment and 10 lb/lb phosphorus removed for the overall plant. As the effluent quality from the treatment plant gradually improved and a nitrifying population became established, chlorine dosage for disinfection was able to be decreased from a level of 6 to 7 mg/l to a level of about 3 mg/l and less.

Sludge Production — Combined Primary and Secondary Sludge

From start-up in March 1974 to January 1975, sludge solids were pumped separately from the primary and secondary clarifiers and mixed prior to entering anaerobic digestion. From March to October 1974, this was done over an 8-hr shift and Table 8-7 shows that the combined sludge averaged 3.6 solids. From November 1974 to January 1975, this was done over a 16-hr shift and the combined solids averaged 2.2%. The additional pumping during the longer operating shift apparently resulted in dilution of the solids. To regain the previously high solids concentration the

secondary solids were returned to the primary clarifier to combine with the primary solids beginning in February 1975. From that point on, 4 to 5% solids were being sent to anaerobic digestion. This is much higher than would be normally attained using an activated sludge process which allows an anaerobic digester to be reduced in size and which also requires less heat to maintain the digester at the desired operating temperature. After digestion, the final sludge content averaged 5.6% of solids, as shown in Table 8-8. This dense final sludge solids content allows a more economical design for a subsequent solids dewatering system.

An average of 0.87 lb of combined primary and secondary sludge produced per pound of BOD was removed by the primary and secondary treatment system. This sludge production includes all suspended solids removed by the treatment plant and all alum sludge produced, as well as the biological sludge generation. Sludge production was lower during the spring and summer months than the fall and winter. It is likely that the warmer wastewater temperature during spring and summer increases the rate of the endogenous respiration by the biological cultures resulting in a lower sludge production.

TABLE 8-8

Sludge Digestion Data for RBC Secondary Treatment Tests at Gladstone, Michigan

Month (1974)	Temperature °F	Digesting sludge Percent solids	Volatile acids, mg/l	pH	Supernatant SS mg/l	BOD mg/l	Digested sludge Percent solids	Percent volatile	Daily gas production ft³ × 100
March	71	–	–	–	–	–	–	–	–
April	71	–	–	–	–	–	–	–	–
May	73	–	228	–	–	–	–	–	–
June	80	6.7	–	–	–	–	6.7	51	50
July	79	6.8	–	–	–	–	–	–	51
August	79	7.0	–	–	–	–	7.0	52	48
September	89	–	–	7.0	933	353	5.7	52	46
October	88	–	264	7.1	743	513	4.8	51	45
November	87	–	85	7.1	390	290	4.1	51	–
December	85	–	66	7.0	620	205	4.8	53	–
(1975)									
January	87	–	243	6.8	620	407	5.3	49	–
February	93	3.1	526	6.8	423	183	–	–	–
March	90	–	430	6.9	700	243	6.4	46	40
April	92	–	701	6.9	1,000	447	5.8	50	44
May	91	–	745	6.9	1,030	380	5.7	50	42
June	91	–	235	7.0	1,126	473	5.3	50	52
July	90	–	441	6.9	2,328	543	5.4	51	58
August	86	–	635	6.9	1,865-	394	5.3	50	56
September	94	–	508	6.8	1,240	457	4.2	48	57

Stage Analysis

On several occasions, samples were taken from each of the stages of treatment in the rotating contactor process. Data from these samples show the progressively increasing degree of treatment provided in the subsequent stages of rotating media. Tables 8-9, 10, and 11 contain stage data collected on three separate occasions. On September 17, 1974, the wastewater flow was 1.0 mgd at a temperature of 63°F. The influent BOD was 123 mg/l and the final effluent after clarification was 11 mg/l. The soluble BOD was slightly less than half of the total BOD, and was progressively reduced to 8 mg/l in the final stage of media. Note, however, that most of the soluble BOD removal occurred within the first three stages of operation. The mixed liquor leaving the sixth and final stage of the process contained 68 mg/l of suspended solids, which was reduced to 14 mg/l after passing through the final clarifier. Comparing this amount of sludge production to the reduction in total BOD by the RBC process indicates a sludge production rate of about 0.5 lb/lb BOD removed by the secondary treatment system.

Samples taken on November 6, 1974, at a reduced hydraulic loading show higher percentage removals of soluble BOD by the upstream stages. The effects of adding alum to the rotating contactor effluent prior to entering the secondary clarifier can be seen by comparing solids levels in Stage 6 with the clarifier effluent. In Stage 6, the suspended solids (after quiescent settling) were just 8 mg/l and were 75% volatile. The clarifier effluent showed a significant increase in total suspended solids but only a modest increase in volatile suspended solids (due to the greater efficiency of quiescent settling). The volatile solids content decreased to less than 50%, indicating an increase in inert solids from chemical addition. Basing a sludge production measurement on the clarifier effluent indicates just 0.28 lb sludge produced per pound of BOD removal, which seems too low. Basing the measurement on Stage 6 indicates 0.40 lb sludge/lb BOD removal which appears to be more realistic. Table 8-10 also shows the gradual increase in mixed liquor D. O. as the wastewater passes from stage to stage.

A similar set of stage samples was taken on January 22, 1975, when the wastewater flow was 0.41 mgd and the temperature was 46°F (see

TABLE 8-9

Secondary Treatment – Stage Analysis

Gladstone, Michigan
September 17, 1974

Flow = 1.0 mgd   Temperature = 63° F

	Total BOD mg/l	Soluble BOD mg/l	COD mg/l	MLSS mg/l	SS* mg/l
Influent	123	53	238	–	84
Stage					
1	–	32	150	–	48
2	–	23	135	–	40
3	–	12	86	–	27
4	–	10	84	–	24
5	–	9	71	–	20
6	–	8	62	68	–
Effluent	11	–	55	–	14

*After 20 min of quiescent settling.

TABLE 8-10

Secondary Treatment – Stage Analysis

Gladstone, Michigan
November 6, 1974

Flow = 0.623 mgd   Temperature = 55° F

	Total BOD mg/l	Soluble BOD mg/l	COD mg/l	MLSS mg/l	SS mg/l	VSS mg/l	D.O. mg/l
Influent	91	44	168	–	51	40	0.9
Stage							
1	–	–	–	–	–	–	1.8
2	41	16	94	43	22	18	2.0
3	–	–	–	–	–	–	2.8
4	20	6.5	58	48	12	8	3.0
5	–	–	–	–	–	–	3.1
6	16	4.5	52	46	8	6	3.4
Effluent	10	7	40	–	23	10	5.2

Table 8-11). To determine the treatment efficiency of the RBC process at higher hydraulic loadings, all of the wastewater flow was put through one half of the rotating media. This operation began on January 7, so the process had been operating under these conditions for approximately 2 weeks. Stage samples indicate that the soluble BOD removal efficiency was maintained and total effluent BOD increased only slightly. Mixed liquor suspended solids in each stage of treatment were less than 45 mg/l. These solids represent a very small fraction of the total amount of biological solids in the system. This points out the inherent stability of the process because any hydraulic shock load would result in a loss of only this small amount of solids, while the vast majority of biological solids in the system remain attached to the rotating surfaces. The low suspended solids concentrations also result in very low solids loading on the final clarifier and permits more flexibility in clarifier design. On January 22, the effluent solids concentration of 43 mg/l was

TABLE 8-11

Secondary Treatment — Stage Analysis

Gladstone, Michigan
January 22, 1975

Flow = 0.41 mgd   Temperature = 46°F

Half of RBC Plant Operating

	Total BOD mg/l	Soluble BOD mg/l	Soluble COD mg/l	MLSS mg/l	SS mg/l
Influent	91	56	116	–	50
Stage					
1	–	30	80	43	31
2	–	–	–	44	–
3	–	15	47	45	17
4	–	12	44	42	14
5	–	8	40	38	11
6	–	8	43	43	8
Effluent	12	–	47	–	10

reduced to 10 mg/l by the final clarifier. Comparing this amount of solids produced to the amount of BOD removed indicates a sludge production rate of 0.42 lb/lb of BOD removed. Note that alum addition was suspended in January 1975, and the effluent suspended solids value was comparable to the value for a settled sample from Stage 6. Also note that the monthly average value for January in Table 8-6 shows improved total suspended solids removal over preceding months.

Nitrification

The Gladstone plant was not designed to achieve nitrification, however, because it often operates in the range of loadings of 1.0 to 2.0 gpd/ft^2, nitrification occurs. Ammonia analyses on weekly grab samples began in July 1974 (see Table 8-12). Influent ammonia values varied from 14 to 34 mg/l, while effluent values through the middle of October were generally about 1.0 mg/l and less. At that point, anaerobic digester operation was changed so that supernatant was returned once each week instead of once each day. A week of accumulated supernatant is a significant load considering that an analysis showed that it contained about 450 mg/l ammonia nitrogen. From mid-October to the end of December, effluent ammonia nitrogen varied from 1.0 mg/l to as high as 10 mg/l. Some of the variation was due to grab sampling which coincided with supernatant return. Sampling during the normal workday when peak

flows occurred did not take into account the lower effluent levels which are produced overnight during lower flow periods. Wastewater temperature also decreased during this period so that all three factors combined to produce variable effluent ammonia nitrogen levels.

It can be seen here that anaerobic digester supernatant return must be more closely controlled on plants designed for nitrification. A storage tank for the supernatant with timer-controlled feed pumps will permit feeding the supernatant to the process on a more continuous basis. A larger portion of the supernatant can also be fed during low flow periods to provide more uniform ammonia loading on the process and maximize nitrification efficiency.

During the test period in January, the three rotating contactor shafts containing nitrifying cultures were shut down. The interruption lasted several weeks and resulted in loss of most of the nitrifiers. From February through May the effluent ammonia nitrogen levels generally varied from 2 to 8 mg/l at hydraulic loadings of 1.0 to 2.0 gpd/ft^2. Because of the low wastewater temperature, the nitrifiers could not grow fast enough to reestablish predominance on the latter stages of media even at the low loading rates. In June, the wastewater temperature rose rapidly to 59°F and a nitrifying culture developed to produce average effluents of 2.2 to 2.9 mg/l of ammonia nitrogen at loading rates of 1.6 to 3

TABLE 8-12

Summary of Nitrification Data

Gladstone, Michigan

Month 1974 and 1975	Average flow mgd	Temperature °F	Ammonia (NH₃ – N) mg/l Influent Range	Average	Effluent Range	Average	Percent removal
July	0.873	62	–	21.0	–	1.0	95.2
August	0.774	65	14.0–34.0	22.4	<1.0	<1.0	96.0
September	0.802	62	16.0–27.0	21.5	1.0–1.4	1.2	94.4
October	0.597	58	17.0–28.0	23.6	<1.0–8.0	3.7	84.3
November	0.676	55	14.0–17.0	15.0	2.0–9.0	4.8	68.0
December	0.563	51	11.0–27.0	15.5	1.0–10.0	3.6	76.8
January (1975)	0.551	47	9.4–22.0	17.3	2.0–16.0	8.5	50.9
February	0.499	47	10.0–22.0	14.0	7.0–24.0	14.0	–
March	0.619	47	5.0–16.0	11.3	2.0–7.0	3.5	69.0
April	0.986	47	12.0–22.0	18.4	4.0–18.0	8.6	53.3
May	0.903	53	14.1–17.0	15.7	4.6–7.6	5.6	35.7
June	1.044	59	6.3–22.0	13.6	1.2–7.3	2.9	78.7
July	0.843	64	13.0–21.0	16.4	1.3–4.7	2.2	86.6
August	0.782	68	5.7–33.0	19.3	1.0–7.3	2.3	88.1
September	0.888	66	11.1–22.8	15.7	1.7–3.4	2.4	84.7

gpd/ft² from June through September. This experience points out an important consideration for plant operation during winter. A serious plant upset during cold wastewater temperature conditions which causes loss of the nitrifying population will prevent a return to previous levels of nitrification efficiency until wastewater temperatures return to summertime values. The extremely low growth rates of the nitrifiers will prevent their regrowth until wastewater temperature rises above 55°F. A treatment plant designed to achieve nitrification must start up about 8 weeks prior to the onset of low wastewater temperatures. Once established, however, a nitrifying growth will maintain activity at low wastewater temperature and at the appropriate loading rate can achieve complete nitrification at temperatures as low as 40°F. These characteristics are likely to be true for all biological treatment processes.

Power Consumption

To confirm power consumption estimates for the rotating contactor process, power consumption measurements were made at the Gladstone plant using a polyphase wattmeter on each drive motor. The electrical power consumption of each shaft is given in Table 8-13. It is apparent that the initial stages of media with thicker growths con-

sume more power than the latter stages with thinner growths. The shafts at Gladstone each contained 20.7 linear ft of media. Scaling these power measurements to the standard maximum shaft length of 25 ft results in an average power consumption of just over 5.0 hp/shaft. This is a slightly conservative estimate of power consumption because the larger shafts will be operating at a greater fraction of the 7.5 hp capacity of the drive motor which will increase its efficiency.

PACKAGE PLANT APPLICATIONS

Subdivision

A wastewater treatment plant incorporating the rotating contactor process was placed into operation in May 1971, in the Blackberry Hill subdivision of Brewster, New York. Table 8-14 contains operating data from this installation for the period of February through September 1972.

The Brewster treatment plant consists of a 24-hr septic tank, rotating biological contactor equipment, a secondary clarifier, and a gravity sand filter. The plant was designed to provide 95% removal of BOD and suspended solids to a flow of 32,000 gal of domestic wastewater per day. This installation uses a rotating contactor assembly in concrete tankage and in a side-by-side configura-

TABLE 8-13

Power Consumption Measurements, Gladstone, Michigan

Shaft #	Stages	hp (20.7 ft of media)	hp Equivalent (25 ft of media)
East side			
1	1 and 2	4.96	6.0
2	3 and 4	3.75	4.5
3	5 and 6	3.5	4.2
West side			
1	1 and 2	5.6	6.8
2	3 and 4	3.7	4.5
3	5 and 6	3.6	4.4
			Average = 5.07

TABLE 8-14

Profile for BOD$_5$* for Brewster, New York Package Plant

Month (1972)	Flow gpd	Temperature °F	Influent to RBC	Stage # 1	2	3	4
	Wastewater values in mg/l						
February	4,500	41	125	21	10	9	6
March	6,800	42	131	15	9	7	6
April	5,900	46	129	14	9	7	7
May	6,000	53	121	12	7	5	6
June	5,600	58	109	15	5	4	3
June	7,100	60	127	9	6	4	4
July	7,300	65	152	12	8	5	5
August	8,000	67	220	17	10	7	6
September	7,800	66	123	11	7	5	5
Profile of NH$_3$-N Values in mg/l							
February	4,500	41	29.9	16.0	5.5	3.8	1.6
March	6,800	42	25.0	13.3	4.5	1.7	1.3
April	5,900	46	26.4	12.5	3.8	0.5	0.1
May	6,000	53	28.3	9.2	0.7	0	0
June	5,600	58	29.5	2.3	0.2	0	0
June	7,100	60	26.8	5.5	0	0	0
July	7,300	65	28.5	6.5	1.8	0.5	0.1
August	8,000	67	33.3	16.4	3.7	1.2	0.8
September	7,800	66	34.0	18.6	5.0	1.3	0.7

*Allylthiourea added to dilution water to suppress nitrification.

Note: August 30, 1972 — 6,000 gal of sludge was removed from septic tank.
Note: Each monthly average consists of two to five samples.

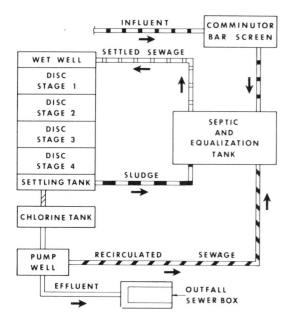

FIGURE 8-16. Rotating contactor package plant flow diagram at Rehrersburg, Pennsylvania. (From Schladitz, C. and Lecker, J., *Water Pollut. Center Assoc. Penn. Mag.*, May–June 1974. With permission.)

tion as described in Chapter 4. Raw wastewater first flows to a septic tank for primary treatment and flow equalization. Effluent from the septic tank then flows to the rotating contactor equipment, consisting of a rotating bucket feed mechanism, four stages of rotating discs, providing a total surface area of 21,000 ft^2, and a secondary clarifier with a rotating sludge scoop. The rotating contactor process portion of the plant was designed for 90% removal of BOD and suspended solids. Effluent from the secondary clarifier flows to the gravity sand filters for polishing treatment and secondary sludge flows back to the septic tank for digestion and storage.

During the operating period reported here, the plant has consistently produced an effluent of excellent quality, both in terms of BOD and ammonia nitrogen removal. The only significant maintenance requirement was the removal of 6,000 gal of sludge from the septic tank, which accumulated during 16 months of operation. The plant operates at just a fraction of its design flow. The RBC produces higher quality effluent when underloaded, unlike activated sludge systems which have difficulty when operated at low loads. Examining the effluent quality from the upstream stages indicates that it will continue to produce an excellent effluent as the flow gradually increases over the years.

Rehabilitation Center

A rotating contactor installation is in operation at the Teen Challenge Training Center, a drug rehabilitation institution, located near the village of Rehrersburg, Pennsylvania.[3] A flow diagram for the plant is shown in Figure 8-16. The wastewater treatment plant contains a comminutor with a by-pass bar screen, a septic and equalization tank for pretreatment, a rotating contactor unit with bucket feed mechanism and rotating scoop-clarifier, and a chlorine contact tank. All components except the pretreatment equipment and septic tank were installed in a single concrete tank and enclosed in a concrete block building. The rotating contactor contains 45,800 ft^2 of media which was selected for a design flow of 20,000 gpd which occurs during a 16-hr period. A small pump was installed in the chlorine contact tank for effluent recycle during holidays and vacations when the rehabilitation center is shut down. This procedure utilizes the soluble organic materials in the septic tank as a food source for the rotating contactor system until wastewater flow is reinstated.

The Rehrersburg treatment plant started operation in December 1973. Test data collected between January and April 1974 are presented in Table 8-15. The values are based on 12- to 14-hr composite samples taken semimonthly. Wastewater temperature was 46°F from start-up to mid-March which accounts for the gradual improvement in BOD and suspended solids removal and the lack of nitrification during the winter months. In April, wastewater temperature rose to 55°F and a nitrifying population quickly developed to produce effluents of less than 1.0 mg/l ammonia nitrogen. A sample taken late in April 1974 showed 70 mg/l of ammonia nitrogen reduced to less than 0.1 mg/l. The warmer wastewater also resulted in less than 10 mg/l BOD and suspended solids in the effluent.

Summer Camp

A rotating biological contactor unit has been in operation at Camp Horseshoe, Tucker County, West Virginia, since 1972. It operates in conjunction with a septic tank as described in Chapter 4 for the over-and-under configuration and as shown in Figure 4-25. The unit contains 4,500 ft^2 of media surface area and is designed to treat 9,000 gpd of wastewater. Figure 8-17 shows the enclosure for the treatment plant which was constructed to be compatible with other buildings

TABLE 8-15

Test Data for Rehresburg, Pennsylvania Package Plant

Sewage flow		Tot. Sus. Solids mg/l		Ammonia Nitrogen mg/l		Phorphorus mg/l		BOD₅ mg/l	
gpd		Raw	Finished	Raw	Finished	Raw	Finished	Raw	Finished
January	6,086	230	29	32	43	12	6	284	21
February	5,675	254	6	50	38	22	13	268	10
March	5,305	319	15	29	22	6	7	346	13
April	4,100	292	6	58	0.29	9	8	345	7.5
Average	5,292	274	14	42	26	12	8.5	311	12.8

From Schladitz, C. and Lecker, J., *Water Pollut. Center Assoc. Penn. Mag.,* May-June 1974. With permission.

FIGURE 8-17   RBC building at Camp Horseshoe, West Virginia.

on the camp ground. Effluent from the treatment system is sprayed into a pine forest for ultimate disposal. Because the camp operates only during the summer months, this provides a convenient means of effluent disposal.

During the summer of 1972, the plant was monitored as part of an EPA research project.[4] A summary of the data is presented in Table 8-16. The data in Table 8-16 represent performance at about half of design flow or a rotating contactor process hydraulic loading of about 1.0 gpd/ft². The performance levels shown are very good for suspended solids removal but not as good as expected for BOD removal. Figure 8-18 shows BOD removal as a function of operating time. During the first week of operation, the plant provided secondary treatment. Percentage BOD removal was lower than expected at a 1.0 gpd/ft² loading rate and was also very erratic. This

behavior is thought to be due to the heavy use of disinfectants during cleaning of showers. During a visit by the author, a distinct medicinal odor was noticed within the treatment plant along with a copious amount of foam within the equalization chamber of the septic tank. The use of large amounts of disinfectants was later confirmed by the camp operator. Operating conditions were such that nitrification should have occurred but did not, which is further evidence that an inhibitory substance was present. However, use of the disinfectants had little effect on suspended solids removal. Figure 8-19 shows excellent removals during the test period.

There is a valve on the clarifier at Camp Horseshoe which allows the final effluent to flow back to the septic tank rather than be discharged. It is meant to be used under zero flow conditions

TABLE 8-16

Performance of Package Plant at Camp Horseshoe, West Virginia

June 18 to July 22, 1972

	Raw (mg/l)	Settled (mg/l)	Effluent (mg/l)	% Rem. RBC	% Rem. Total
BOD	305.8	151.5	32.8	78.3	89.3
COD	494.4	249.1	92.7	67.8	81.2
TOC	113.2	73.5	26.6	63.8	76.5
SS	393.8	49.8	12.4	75.1	96.9

July 23 to August 24, 1972

	Raw (mg/l)	Settled (mg/l)	Effluent (mg/l)	% Rem. RBC	% Rem. Total
BOD	213.3	252.8	31.8	87.4	85.1
COD	615.0	559.5	144.0	74.3	76.6
TOC	122.7	124.3	33.2	73.3	72.9
SS	251.5	11.8	2.8	76.3	98.9

NOTE: BOD time periods are June 18 to July 15, 1972, and July 16 to August 24, 1972.

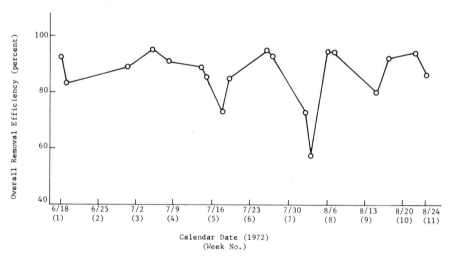

FIGURE 8-18. BOD removal efficiency at Camp Horseshoe, West Virginia.

to permit recirculation of septic tank contents and provide some sustaining organic material to the attached biomass. Zero flow conditions exist each weekend at the camp, however, the recirculation feature was not used. This could also have contributed to the irregular BOD removal efficiency.

Maintenance and operator attention required for sewage treatment plants in remote areas is of great concern. Since a full-time or trained operator is seldom available, these factors must be carefully considered when selecting the process to be used. The Camp Horseshoe plant operated continuously for approximately 11 weeks during the EPA study mentioned above. During this period, a total of only 14 hr, or an average of 1.3 hr/week, of maintenance time was required to keep the plant running. Repair time consisted of cleaning and drying contacts in the buffer tank pump float controls (3 times), and tightening the main shaft sprocket chain (1 time) for a total of 6 hr. The

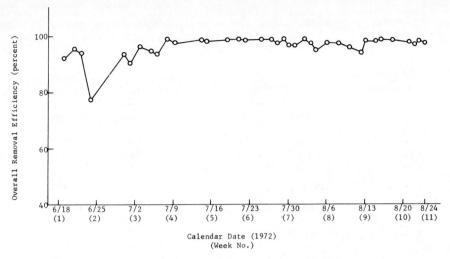

FIGURE 8-19. Suspended solids removal efficiency at Camp Horseshoe, West Virginia.

balance of operator attention consisted of preparing the chlorine solution 3 times a week for a total time of 45 min/week. The operator reported that start-up and shutdown procedures required a total of 5 hr. Shutdown procedure for the system consists simply of draining the process tankage and allowing the attached biomass to dry; it is not necessary to remove the biomass. At start-up the following season the dry biomass provides a substrate for rapid development of a new biomass.

At the end of the study, sludge accumulation in the septic tank was estimated at 440 gal/10^6 gal of wastewater treated. This is low enough to allow removal of the accumulated solids for ultimate disposal just once at the end of the summer.

INDUSTRIAL WASTE TREATMENT

Small Cheese Factory

A rotating contactor plant has been in operation at the Eiler's Cheese Factory in DePere, Wisconsin since 1970.[5] The Eiler Cheese Co. processes 30,000 lb of milk into 3,000 lb of American, Cheddar, and Colby cheeses daily. About half of the milk received is in cans and the washings from these cans constitute a major source of the wastewater; the balance is whey washwater. A total daily flow of 3,000 gal of wastewater results from the cheese factory. Peak daily flows of up to 5,400 gpd are experienced at the plant during peak production seasons. All whey waste is sold to local farmers, and used as an animal food.

The RBC facility is located approximately 300 ft from the owner's residence and 500 ft from the cheese plant. Domestic waste from 11 people is added to the dairy wastewater.

The treatment facility consists of three major subsystems: 1. A septic pretreatment and flow equalization unit, 2. a four-stage RBC unit with secondary clarifier and, 3. a 157,000-gal, 30-day polishing lagoon. The facility was designed on the basis of design data from a similar installation in Switzerland.[6] Design specifications are presented in Table 8-17. A flow schematic of the facility is shown in Figure 8-20. Raw waste flows first to the septic tank unit which serves several purposes. First, it is a collection basin for primary and secondary solids, second, the compartmental chambers serve as flow equalization units, reducing the impact of peak hydraulic and organic loads, and third, it provides sludge digestion and storage.

The RBC unit and associated items are situated in a vault 12 ft below ground level. Electrical utilities were installed on one wall of the vault for easy maintenance and servicing. In accordance with regulatory agency requirements, a chlorinator is provided for effluent disinfection. The rotating contactor unit is essentially identical to that shown in Figure 4-19.

Wastewater flows from the septic tank system to the RBC unit. A bucket pump attached to the RBC shaft pumps the wastewater from the feed chamber into the first stage of media.

The RBC unit is composed of four stages in series. Each stage contains 22 molded polystyrene

TABLE 8-17

Dairy Plant Design Specifications

BOD removal requirements	BOD mg/l		lb BOD/day		Overall reduction
	Inlet	Outlet	Inlet	Outlet	Percent
Septic tanks	2,000	1,000	90	45	50
RBC, four stages	1,000	100	45	4.5	95
Polishing lagoon	100	50	4.5	2.3	98

Flow, gpd – 3,000 gal/day (average)

5,400 gal/day (maximum)

Organic load – 2,000 mg/l BOD or less

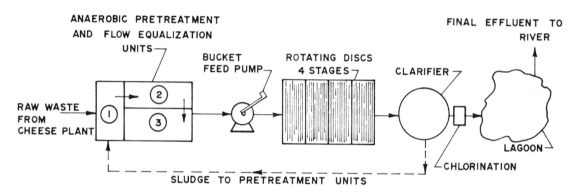

FIGURE 8-20.   Process flow diagram for Eiler's Cheese Factory.

discs, 10 ft in diameter. Total surface area is 13,800 ft². Discs are rotated at a peripheral velocity of 62 ft/min, with wastewater passing from stage to stage through openings in the cross-tank bulkheads. Operation of the RBC septic tank system is essentially the same as described for Figure 4-24.

The polishing lagoon following the RBC plant is a 30-day, facultative detention pond. Surface area is approximately 60 × 110 ft with a capacity of 157,000 gal. Nominal depth is 4.5 ft. The basin is situated about 100 ft from the rotating contactor vault. Lagoon effluent is discharged to a small field stream which eventually enters the East River leading to the Fox River.

The below ground construction of the treatment facility resulted in appreciable dampening of the diurnal and seasonal temperature variations in the Green Bay/DePere, Wisconsin area. Figure 8-21 compares U.S. National Weather Service temperatures for the area with the recorded mixed liquor

temperature and the recorded RBC vault air temperature. During the long period of subfreezing ambient temperatures, temperatures of the air in the vault and of the mixed liquor were on the average 42° F and 52° F, respectively, and never approached freezing. During mid-winter months, mixed liquor temperatures were only 20° to 25° lower than those observed during mid-summer conditions, indicating that sub-surface vault construction provides good temperature stability without addition of supplementary heat.

Operating hours of the cheese plant are from 4:00 a.m. to 5:00 p.m. Virtually all the wastewater flows to the facility during this time. The highest hydraulic load occurs between 2:00 p.m. and 4:00 p.m. due to periods of washdown and cleanup following the cheese-making operations. Figure 8-22 shows a typical daily flow pattern to the rotating contactor unit. The pattern was determined from continuous measurements of the wastewater level in the feed well. Although waste-

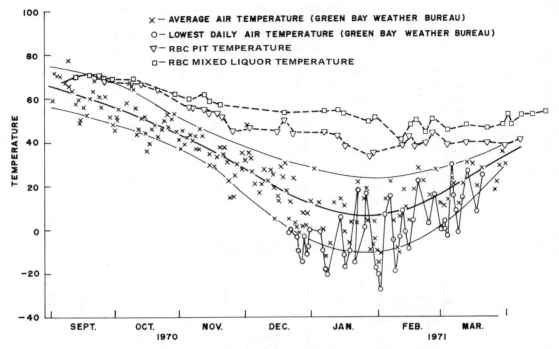

FIGURE 8-21   Temperature variations with below ground construction.

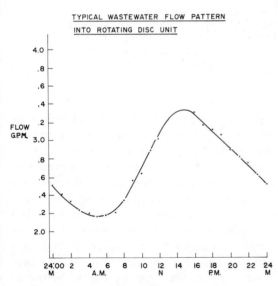

FIGURE 8-22.   Wastewater flow pattern with flow equalization and bucket pump mechanism.

water flow to the treatment plant was very irregular, the combination of flow equalization tank and bucket feed mechanism resulted in a flow

pattern to the RBC process where the peak-to-average flow ratio is just 1.2:1.0. Table 8-18 summarizes system performance data accumulated during the period of April 19–28, 1971. An average influent raw waste BOD of 1,062 mg/l and an average RBC effluent BOD of 42 mg/l were observed. A RBC effluent BOD of 100 mg/l or less, or a 95% reduction of influent raw waste BOD up to 2,000 mg/l were specified. As table 8-18 indicates, both of these conditions were met.

Operating and maintenance costs of Eiler's Cheese Co. for the RBC installation were very low. In general, only a daily visual inspection of the facility was required. Lubrication of motors and bearings requires approximately 1 man-hr/month. Annual power cost for the facility, at a rate of $80/hp-year, is approximately $100. An annual removal of accumulated solids from the septic tanks and miscellaneous other material purchases costs about $110. The total annual operating cost is thus about $210.

Large Dairy Products Company

There is a rotating contactor process installa-

TABLE 8-18

RBC Dairy Waste Performance Data

Date	BOD, mg/l		Percent BOD reduction	Lagoon effluent BOD, mg/l
	Raw	Final clarifier effluent		
April 19, 71	840	40	95	17
April 20, 71	720	32	96	27
April 21, 71	780	21	97	32
April 22, 71	1,700	29	98	22
April 23, 71	1,240	54	96	54
April 24, 71	1,100	65	94	53
April 25, 71	705	53	92	50
April 26, 71	840	30	96	48
April 27, 71	1,540	48	97	46
April 28, 71	1,155	48	96	17
Average	1,062	42	96	36

DAIRY WASTE TREATMENT

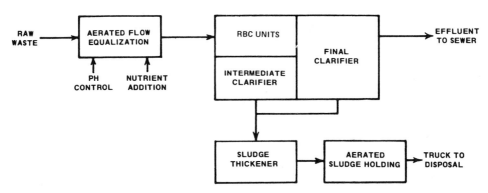

FIGURE 8-23.   Process flow diagram for Zausner Foods.

tion at the Zausner Foods Corp. in New Holland, Pennsylvania, which is providing pretreatment prior to sewer discharge.[7] Wastewater produced at this plant is from bulk receiving of milk and from the manufacture of cottage cheese, sour cream, pudding, and other speciality dairy products. The treatment plant was designed for a flow of 240,000 gpd and a BOD concentration of 2,270 mg/l. Effluent requirement is 227 mg/l BOD at design flow.

Figure 8-23 is a flow diagram of the Zausner Foods Corporation installation. Raw waste flows to existing aerated holding tanks of 100,000-gal total capacity which were converted to flow equalization tanks. Nutrient addition and pH control were also installed at this point. Wastewater is pumped to 12 shafts of media arranged in

2 parallel flow paths of 6 shafts each. Intermediate and final clarification are incorporated into the same tankage as the rotating contactor units. An extension of the tankage provides secondary clarification to minimize overall construction cost. The enclosure for the rotating contactor process also covers the intermediate and final clarifiers. This configuration permits easy expansion of the facility to accept higher wastewater loads or meet higher treatment requirements by installing additional shafts in the secondary clarifier portion of the tankage and constructing a new final clarifier.

Shortly after the plant described above started, it became apparent that the waste needed some form of pretreatment because of the large amount of floating solids. These floating solids were primarily hexane-solubles from pudding manu-

BAKERY WASTE TREATMENT

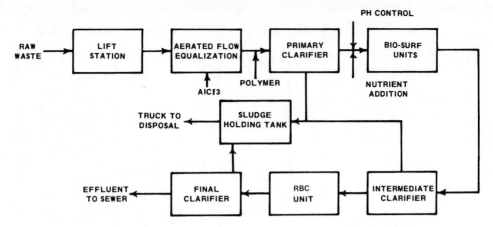

FIGURE 8-24. Flow diagram for S. B. Thomas Bakery waste treatment.

facturing. To eliminate this problem the sludge thickener was converted to a primary clarifier. In the future, this change also will permit chemical pretreatment if loadings increase or treatment requirements become more stringent. Since rotating contactor process sludge comes out of the clarifier at 2 to 3% solids, further thickening is not required. The increased volume of sludge to be trucked to disposal was a more economical alternative than constructing a new primary clarifier.

Waste sludge is trucked to the local municipal treatment plant. Many municipalities which require that industrial wastes be pretreated before being discharged to the sewer system will accept sludge from that pretreatment when hauled separately to the municipal treatment plant. This apparently is the case when biological treatment facilities at the municipal plant cannot handle the organic load from the industry, but whose sludge-handling facilities have sufficient capacity for the biological sludge. Wastewater flows to this pretreatment facility have averaged about 200,000 gpd. The discharge of hexane-solubles from pudding processing has been minimized through inplant procedures. Those remaining are being removed through primary treatment. This has resulted in an influent BOD concentration of under 3,000 mg/l and an effluent discharge to the municipal sewer in the range of 100 to 200 mg/l BOD.

Bakery

A rotating contactor process installation has been in operation at the S. B. Thomas Bakery in Totowa, New Jersey, since 1970.[7] The bakery produces English muffins, special breads, and other specialty bakery products. The wastewater primarily contains sugar, starch, and grease from pan washing. The treatment plant was designed for a wastewater flow of 50,000 gpd and a BOD concentration of 2,000 mg/l after primary treatment. Discharge to the municipal sewer system must be 300 mg/l BOD or less.

Figure 8-24 is a process flow diagram for this pretreatment application. Raw wastewater is pumped to a 50,000-gal equalization tank. From there it flows at a controlled rate to a primary clarifier for removal of grease and settleable matter. A sensor measures pH of the primary effluent and sodium hydroxide is added to the equalization tank to control acidity. Nutrient salts are added prior to entering the rotating contactor units. The initial RBC units consist of two shafts of media installed in prefabricated steel tankage. The shafts are set up in parallel with wastewater flow parallel to the shafts and with each shaft divided into two stages by cross-tank bulkheads. Effluent from the first units flows into an intermediate clarifier for separation of biological solids prior to entering the final RBC. This unit is divided into four stages. Effluent from the last unit receives final clarification and then is discharged to the municipal sewer. Sludge from the three clarifiers flows to a holding tank and is periodically trucked to disposal.

Shortly after this plant started, BOD loads were well in excess of those anticipated in design. BOD concentration after primary settling was in the

WINERY WASTE TREATMENT

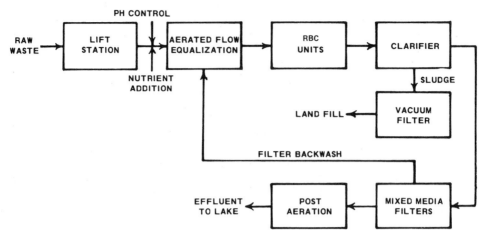

FIGURE 8-25. Flow diagram for Gold Seal Winery waste treatment.

range of 3,000 to 5,000 mg/l. Although wastewater flow was just 30,000 to 35,000 gpd, the total organic load resulted in effluent BOD concentrations of 600 to 1,200 mg/l, well above the requirement. To help alleviate the overloading problem, inplant operations and cleanup procedures were watched closely to avoid unnecessary discharges of organic materials into the plant sewers. These steps reduced the BOD concentration close to the 2,000 mg/l-design condition and the treatment plant was then able to produce an effluent of less than 300 mg/l. However, those procedures were effective for only a short time before wastewater strengths again exceeded design conditions.

As a more permanent solution to the problem of organic overload, it was decided to utilize chemical additions in the primary clarifier to reduce the BOD concentration of the primary effluent. Initially, aluminum chloride alone was added and achieved about 30% BOD reduction. Then a polymer was also added and together they achieved a 50% BOD reduction which brought the primary effluent well within the 2,000 mg/l-BOD limit. With the primary effluent in the range of 900 to 1,600 mg/l BOD, the effluent to the municipal sewer was 100 to 200 mg/l.

The power consumed by this installation was under 5 hp. This scales up to a power consumption of about 80 hp/mgd capacity.

Winery

A 350,000-gpd rotating contactor installation is in operation at the Gold Seal Winery in Hammondsport, New York.[7] This project started with a pilot plant program which evaluated the process along with activated sludge and extended aeration. The results of this pilot plant program were reported by LaBella and Thacker.[8] BOD removal efficiency demonstrated during the pilot plant study is shown in Figure 6-19.

The level of treatment required for full-scale design is 95% BOD reduction to yield a secondary effluent of under 35 mg/l BOD. To achieve further BOD removal the effluent is passed through mixed media filters and then receives postaeration prior to being discharged to a lake. Figure 8-25 shows the process flow scheme. Aerated flow equalization is utilized along with pH control and nutrient addition. Intermediate clarification was not used for the process because of the relatively low 700 mg/l-BOD concentration. Rotating contactor process equipment for this installation consists of nine shafts of media arranged in a single flow path. This was done to simplify construction on the side of a steep hill. Wastewater is distributed equally to the first three shafts via a system of troughs and weirs to reduce organic loading on the initial stages of treatment. After treatment by the first three shafts the wastewater flows through the remaining six shafts in series with flow perpendicular to the shafts. This plant has undergone a thorough evaluation during several pressing seasons and has performed as designed. Figure 8-26 shows the RBC portion of the plant which is covered by an inflated vinyl building.

Vacuum filtration of undigested biological solids for the final clarifier has proved difficult. As

FIGURE 8-26. Inflated vinyl building for RBC installation at Gold Seal Vineyards.

an alternative, some of the sludge solids are being plowed into the ground in the vineyards, which also provides a source of nutrients and conditioning of the soil.

Distillery

The U.S. EPA awarded a demonstration grant to the American Distilling Co. in Pekin, Illinois, to test the rotating contactor process on distillery wastes.[7] RBC process equipment was installed to treat a portion of the 0.5 mgd of wastewater from the distillery. Figure 8-27 shows the process flow diagram for this installation. Partial cooling and some pH adjustment of the wastewater were performed before entering the treatment plant. Nutrients were added at the lift station, and flow equalization was utilized after grit removal. Flow equalization also helped reduce variations in BOD concentration, pH, and temperature, so that the wastewater entering the rotating contactor process was generally less than 90°F.

Two RBC units were used for this demonstra-

tion plant described above. The first was divided into two stages and was followed by an intermediate clarifier. The second contained four stages to yield a total of six stages of treatment. During the initial phases of the test program, the equalization tank was unaerated. With retention times in this tank ranging from 8 to 24 hr, septic conditions developed and decreased the efficiency of the subsequent biological treatment. To alleviate this problem, aeration equipment was installed in the equalization tank, which aerated it at a rate of about 100 CFM/lb BOD load. This kept the tank aerobic but resulted in the development of a finely divided activated sludge floc, which would not settle in subsequent clarification steps. To avoid development of this floc, it is recommended that aeration be kept under 50 CFM/lb BOD load. For applications where the efficiency of mixing would not be adversely affected, this should be reduced to 30 CFM/lb BOD load. Figure 6-19 shows average operating data for the test periods when the equalization tank was aerated.

DISTILLERY WASTE TREATMENT

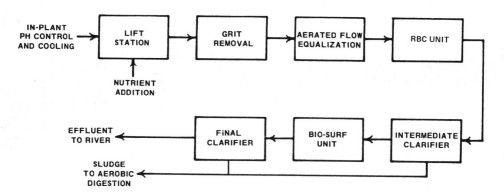

FIGURE 8-27.   Flow diagram for distillery waste treatment.

Poultry Processing

In March 1972, a rotating contactor installation began operation at the Kralis Poultry Co. in Olney, Illinois.[7] Kralis Poultry Co. is a pre-treatment application where the RBC process was designed for 80% BOD removal on a wastewater flow of 130,000 gpd. Wastewater is generated from slaughtering chickens and contains 4,000 to 5,000 mg/l BOD. After primary treatment in an existing Imhoff tank and an air flotation unit, the BOD is reduced to 1,500 to 3,000 mg/l.

Prior to air flotation, the grease content measured as hexane-solubles is 500 to 800 mg/l. This is reduced to less than 200 mg/l to avoid problems in the biological treatment step. Pilot plant experience at another poultry processing plant has shown that hexane-solubles of less than 200 mg/l can be tolerated by the rotating contactor process and will be removed in approximately the same proportion as the BOD content.

The RBC process equipment consists of two shafts, each divided into two stages, operated in series with wastewater flow parallel to the shafts. Figure 8-28 is a process flow diagram for the installation.

Nylon Manufacturing

The first rotating biological contactor process application in the textile industry has been in operation since 1974 at the E.I. Du Pont Co. in Chattanooga, Tennessee.[9] The wastewater is generated from nylon manufacturing and generally contains less than 50 mg/l BOD and less than 10 mg/l ammonia nitrogen. The plant contains 24 shafts of media in an arrangement similar to

Configuration #4 with a common influent channel as shown in Chapter 7. The plant is designed to treat a flow of 3.7 mgd principally for nitrification prior to discharge to the Tennessee River. Pretreatment for solids removal is not necessary on this wastewater, however, the influent is carefully monitored for pH and for components which could be potentially toxic to the nitrifying bacteria. If influent wastewater does not meet specified requirements it is diverted into a one million-gal capacity impoundment pond. Here it can be adjusted before being fed back into the influent flow under controlled conditions. Secondary sludge is aerobically digested using submerged turbine aerators and sent directly to covered drying beds. After drying it is trucked away to a land fill site.

The treatment plant has been consistently producing an effluent of less than 10 mg/l BOD and less than 1.0 mg/l ammonia nitrogen which exceeds the required levels of treatment for discharge to the Tennessee River.

Pulp and Paper Manufacturing

There are presently two rotating contactor applications in the pulp and paper industry. One has been in operation since 1974 at the Northwest Paper Co. in Brainerd, Minnesota, which was designed to treat 2.8 mgd of wastewater from fine paper manufacturing. Because no pulping is done at this mill, the wastewater consists entirely of materials from fine paper manufacturing which results in high concentrations of starch and clay. A significant portion of the clay forms a bond with the starch and passes through a primary settling

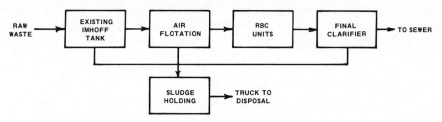

FIGURE 8-28. Flow diagram for Kralis poultry processing waste.

FIGURE 8-29. Interior of RBC building at Boise Cascade Pulp and Paper Co. ·

tank. However, when the wastewater enters biological treatment the clay—starch bond is broken and the clay settles to the bottom of the rotating contactor tankage. These deposits were effectively removed by adding several extensions to the rotating contactor assemblies to help mix the tank and keep the clay particles in suspension.

The RBC equipment consists of 9 rotating contactors arranged in 3 parallel paths of three shafts each and are designed to obtain 80% BOD reduction for wastewater strengths in the range of 150 to 300 mg/l.

The second RBC plant for pulp and paper waste treatment is under construction at the Boise Cascade Company in International Falls, Minnesota. It is planned to begin operation in the fall of 1975. The installation is designed to treat 30 mgd of wastewater from integrated kraft mill and insulation board mill operations. Treatment requirements are 80% reduction at a BOD concentration of about 250 mg/l. The plant consists of 96 rotating contactor units as shown in Figure 8-29 and are arranged as shown in Figure 7-7. In this case, however, the flow pattern is reversed, i.e., wastewater flows from the outside through four stages of treatment and into a common effluent channel in the center of the structure.

During pilot plant testing at Boise Cascade a fine screen was substituted for primary clarification. Fine screens are being used as a substitute for primary clarification on the full-scale plant and the existing primary clarifier was converted to use as a secondary clarifier, which resulted in a significant reduction in construction cost. The rotating biological contactor process was chosen for this application over lagoons and activated sludge processes because of its lower capital and operating costs, simple operation, and its stability under shock loads and excellent toxicity reduction demonstrated during the pilot plant tests.[10-12] When this installation goes into operation late in 1975 it will be the largest rotating biological contactor installation in the world.

REFERENCES

1. **Antonie, R., Kluge, D., and Mielke, J.,** Evaluation of a rotating disc wastewater treatment plant, *Water Pollut. Control Fed. J.,* 46(3), 498, 1974.
2. **Malhotra, S. K., Williams, T. C., and Morley, W. L.,** Performance of a Bio-Disc Plant in a Northern Community, paper presented at the 48th Annu. Conf. Water Pollut. Control Fed., October 5–10, 1975, Miami Beach, Fla.
3. **Schladitz, C. and Lecker, J.,** Teen challenge training center "BIO-SURF" wastewater treatment plant, *Water Pollut. Center Assoc. Penn. Mag.,* May–June 1974.
4. **Sack, W. and Phillips, S.,** Evaluation of the BIO- Disc treatment process for summer camp application, Environ. Protect. Technol. Services, EPA 67012–73–022, August 1973.
5. **Birks, C. W. and Hynek, R. J.,** Treatment of Cheese Processing Wastes by the BIO-DISC Process, Proc. 26th Annu. Purdue Ind. Waste Conf., May 4–6, 1971, W. Lafayette, Ind., 89.
6. **Bretscher, U.,** Small Sewage Treatment Plant with RBC's for The Purification of Dairy Waste Water, Rep. No. 9614, Society of Swiss Sewage Specialists, November 1967.
7. **Antonie, R. and Hynek, R.,** Operating Experience with BIO-SURF Process Treatment of Food Processing Wastes, Proc. 28th Purdue Ind. Waste Conf., May 1–3, 1973, W. Lafayette, Ind., 849.
8. **LaBella, S., Thacker, I., and Teehan, J.,** Treatment of Winery Wastes by Aerated Lagoon, Activated Sludge Process, and Rotating Biological Contractors, or RBC's, Proc. 27th Annu. Purdue Ind. Waste Conf., May 1–4, 1972, W. Lafayette, Ind.
9. BIO-DISC System Shrinks BOD 90% in Prototype CPI Installation, *Chem. Process. Lond.,* July, 70, 1975.
10. **Summer, R. and Bennett, D.,** Effluent treatment by rotating biological surface, *Pap. Trade J.,* October 8, 1973.
11. **Bennett, D., Needham, T., and Summer, R.,** Pilot application of the rotating biological surface concept for secondary treatment of insulating board mill effluents, *Tappi,* 56(12), 41, 1973.
12. **Summer, R. and Bennett, D.,** Pilot Treatability Study of Pulp, Paper and Fiberboard Effluents by Rotating Biological Surface, Proc. Can. Pulp Pap. Assoc. 1973, Pap. Ind. Air Stream Improvement Conf., September 11–13, 1973, St. Andrews, N. B., 89.

Chapter 9

CAPITAL AND OPERATING COSTS

INTRODUCTION

This chapter presents capital and operating costs for the rotating biological contactor process over a wide range of plant sizes and equipment configurations. A detailed comparison of operating costs and a differential capital cost analysis will be made with the activated sludge process.

CAPITAL COSTS

Construction costs for the RBC process are presented in Figures 9-1, 2, and 3 for the equipment configurations described in Chapter 7. The costs were determined by adding the cost of RBC equipment, concrete tankage at $200/yd^3, fiberglass enclosures, freight for RBC units and covers to an average point within the contiguous U.S., and installation costs for cranes, millwrights, electricians, etc. The total installed costs are expressed per unit of wastewater flow treated and shown as a function of hydraulic loading. An estimate for total installed cost for an RBC system can then be calculated for any application using Figures 9-1, 2, or 3 and the appropriate design

curve from Chapters 4, 5, or 6. The cost per unit flow for the specific hydraulic loading from Figures 9-1, 2, or 3 is multiplied by the design flow to arrive at total cost. The costs shown do not include the expenses of pretreatment, secondary clarification, and sludge disposal. They will vary significantly from application to application, depending upon design flow, wastewater characteristics, degree of treatment, site conditions, and existing facilities. These additional costs can be estimated from any of the references pertaining to the subject.[1-3]

RBC equipment costs represent 65 to 70% of the total costs in Figures 9-1, 2, and 3, which indicates that the process is equipment intensive compared to the activated sludge process where the cost of structures and piping constitute the majority of the total installed cost.

Experience has shown that the costs shown in Figures 9-1, 2, and 3 represent 30 to 35% of the total treatment plant cost for plants designed for standard secondary treatment. With plants designed for both BOD removal and nitrification, the RBC portion of the total plant cost increases to 35 to 45%.

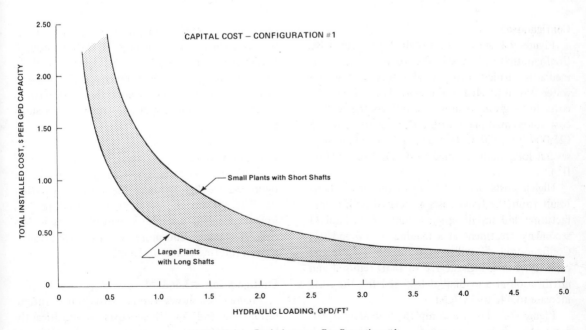

FIGURE 1.   Capital cost – Configuration #1.

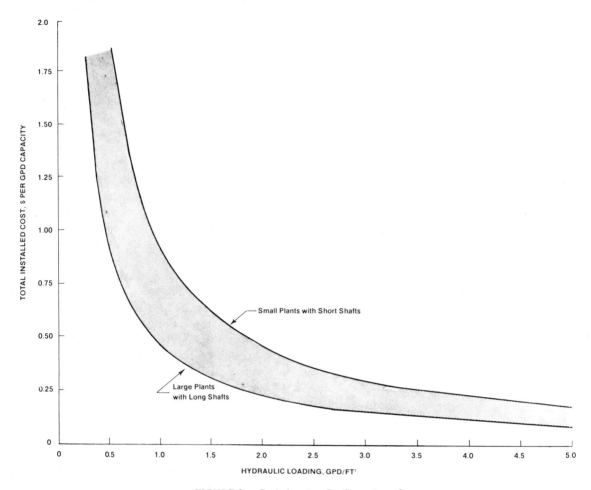

FIGURE 2. Capital cost — Configuration #2.

Configuration #1

Figure 9-1 shows the installed cost for RBC Configuration #1 where the rotating contactor media is divided into several stages, and wastewater flow is parallel to the shaft. The range of the costs for a given loading rate reflects the relative cost for a small plant with a short shaft of media (25,000 to 50,000 ft²) and a large plant with several long shafts of media (100,000 to 300,000 ft²).

Higher costs per unit flow for the small plants result from the fixed costs associated with manufacturing and installing small units. For standard secondary treatment at a loading of 3.0 gpd/ft², Configuration #1 will cost from 19¢ to 39¢/gpd capacity. For a high degree of BOD removal and nitrification at a loading of 1.25 gpd/ft², the costs increase to 44¢ to 98¢/gpd.

Figure 7-3 shows a complete secondary treatment plant with common wall construction, which

incorporates RBC Configuration #1. This plant was constructed for the city of Battle Ground, Washington, in 1974 for $261,000. With a capacity of 0.65 mgd the total cost is just 40c/gpd capacity, which demonstrates the cost savings potential of common wall construction in a small plant.

Configuration #2

Construction costs for Configuration #2 in figure 9-2 are slightly lower than for Configuration #1. This reflects the use of two-stage rather than four-stage media assemblies, which results in less concrete tankage and more media surface area for a given rotating contactor length.

Configurations #3 and #4

Figure 9-3 shows capital costs for Configurations #3 and #4. These costs are significantly lower than for Configurations #1 and #2. This is

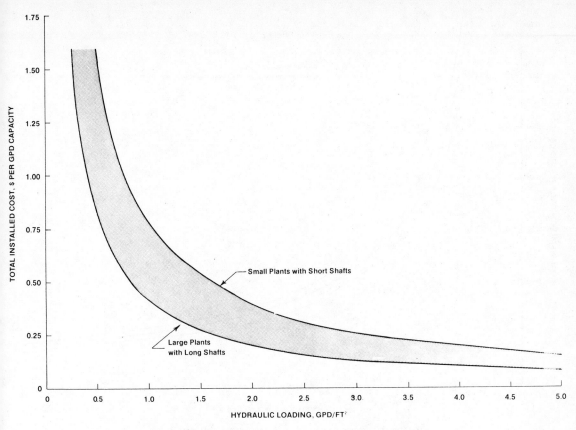

FIGURE 3. Capital cost — Configurations #3 and 4.

the result of the use of flat-bottom tankage with wastewater flow perpendicular to the shaft. Shaft assemblies are completely filled with media, and concrete requirements are reduced from 40 to 50 yd³ for the trapezoidal-shaped tanks to 25 yd³ for the flat-bottom construction. These factors reduce construction costs to 13¢ to 27¢/gpd for standard secondary treatment, and to 33¢ to 63¢/gpd for a high degree of BOD removal and nitrification.

Configurations #3 and #4 use four shaft assemblies in series to achieve four-stage operation. In almost all cases, the RBC units will be maximum length so that the lower limit of the costs in Figure 9-3 is most pertinent.

The use of 25 yd³ of concrete per shaft is meant to reflect average conditions. Depending on soil conditions and piling requirements, this number can range from 18 to 50 yd³/unit.

Figures 9-1, 2, and 3 can also be used to estimate costs for denitrification applications. The RBC units and drive systems will be very similar to normal units, so the costs will also be very similar. The tankage requirements will increase to accom-modate the completely submerged media, but enclosure costs will decrease. Other costs will remain essentially the same, so that total installed costs for standard units will be close enough to use for estimating purposes.

High Density Media

For applications where nitrification is required or when treating dilute BOD concentrations (less than 25 mg/l total BOD or 12 mg/l soluble BOD), the biomass thickness will be reduced so that a higher surface density media can be used. This provides up to 150,000 ft² of area per shaft assembly compared to standard media at 100,000 ft² per shaft, which consequently has a significant effect on total installed cost. Figure 9-4 compares capital cost for several media combinations. For standard secondary treatment of domestic waste or treatment of concentrated industrial wastes, all standard media with a surface density of 37 ft²/ft³ is used, while for combined BOD removal and nitrification of domestic waste or for treat-ment of dilute industrial wastes, a combination of

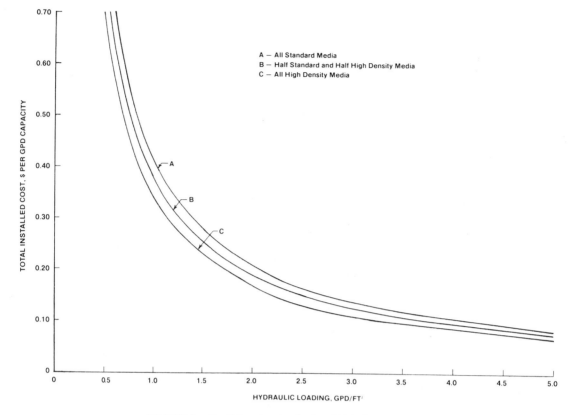

FIGURE 4. Capital cost — standard and high density media.

standard media in initial stages and high density media at 56 ft²/ft³ in later stages can be used. When using half standard and half high density media, the total installed cost is reduced by approximately 10% as compared to all standard media. In addition to having the unit surface area cost of the rotating biological contactor unit decrease with the high density media, further savings are realized by a proportional reduction in the number of tanks, covers, and the shipping and installation costs.

All high density media can be used for application to nitrification of secondary effluents or the treatment of very dilute BOD concentrations. This results in a total cost reduction of about 20%. The cost data in Figure 9-4 incorporate a 20% increase in concrete tankage for high density media to increase the tank volume and maintain the optimum tank volume to media surface area ratio of 0.12 gal/ft².

Power consumption per unit surface area with the high density media is the same as standard media. Maintenance requirements, however, will be reduced because of the reduction in the number of mechanical components.

Each shaft of standard media occupies 450 ft² of area, and each high density media occupies 480 ft² of area including walkways. Therefore, 90 to 100 shaft assemblies can be placed on about 1 acre of land.

Package Plants

Equipment costs for package plant units in steel tankage are shown in Figure 9-5. The units contain either 2.0-m- or 3.2-m-diameter media in a configuration as shown in Figure 4-20. These units operate in conjunction with aerobic or septic tank pretreatment and sludge disposal as shown in Figures 4-21, 25, and 26, so that additional costs are required for a complete installation. However, these additional components are of simple construction with few mechanical parts. Total costs will vary depending upon local soil conditions and architectural requirements.

The costs in Figure 9-5 are expressed per unit of media surface area and are shown as a function of the amount of surface area in the unit. The costs decrease rapidly as more area is supplied, because only the length or diameter of the unit changes with little change in the number and type

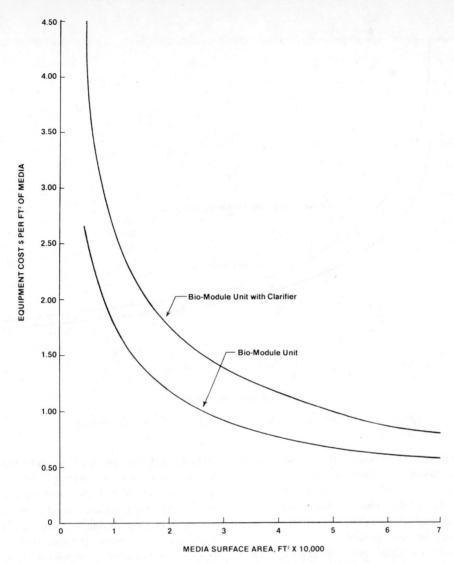

FIGURE 5.   Capital cost – package plants, including steel tankage.

of the other components. The same is true for the clarifier-chlorine contact unit. For some applications, an RBC unit can be used with a separate, conventional clarifier.

Figure 9-6 shows equipment costs for media assemblies and clarifier mechanisms operated as shown in Figure 4-27 and Figure 7-9. These units also have additional costs required for pretreatment and solids disposal. Concrete tankage costs for the media and clarifier mechanism are not included in Figure 9-6. These tanks will often be formed as part of the pretreatment and solids handling system.

The costs shown here for RBC package plant applications are somewhat higher than for activated sludge systems – extended aeration and contact stabilization. This will often be true when treatment levels of 85% BOD removal or less are acceptable and when high power consumption is not a deterrent. However, for applications requiring a consistently high degree of treatment with low operating and maintenance requirements, the rotating contactor system costs will often be lower.

POWER CONSUMPTION

Conservation of energy in wastewater treatment is important to reduce operating costs and to minimize the adverse effect energy consumption

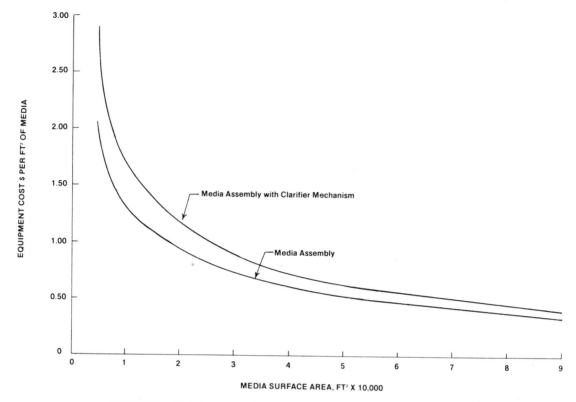

FIGURE 6. Capital cost – package plants, for installation in concrete tankage.

has on the environment. Consumption of electrical energy by a wastewater treatment plant is often accompanied by air pollution and thermal discharges at the power plant. With fossil fuel supplies dwindling, there is further incentive to conserve energy wherever possible.

Power costs are rising and are expected to continue in the same manner. These rising costs, together with the possibility of progressive rate schedules for large users, will have a significant effect on the operating cost of wastewater treatment plants. Because municipal treatment plants are eligible for public funds for treatment plant construction and generally receive nothing for operating costs, the operating costs often outweigh the importance of amortized construction costs in determining total treatment costs for the municipality. This points out the importance of selecting wastewater treatment technology which minimizes power consumption.

Efficient aeration and thorough mixing of tank contents are achieved with low rotational speeds with the rotating contactor process. This results in low power consumption. Figure 9-7 shows power consumption as a function of design hydraulic

loading, and can be used to determine power consumption for most rotating contactor process applications. Installed horsepower will generally be 30% greater than the consumed power to insure proper operation during start-up, and for occasional load imbalances. The indicated power consumption is based on a peripheral media velocity of 60 ft/min for all stages of rotating surfaces in a system.

Power consumption by the process for various degrees of treatment for both BOD removal and nitrification on domestic waste is shown in Figure 9-8. Power consumption for nitrification of secondary effluent is shown in Figure 9-9.

When evaluating secondary treatment systems, total plant costs, both capital and operating, must be considered. Two methods of evaluating power cost savings are described below.

Present Value Analysis

A present value analysis takes a flow of future annual savings and discounts them back to the present time at a specific interest rate to determine their present worth. It is a technique often used by financial analysts to evaluate competing projects.

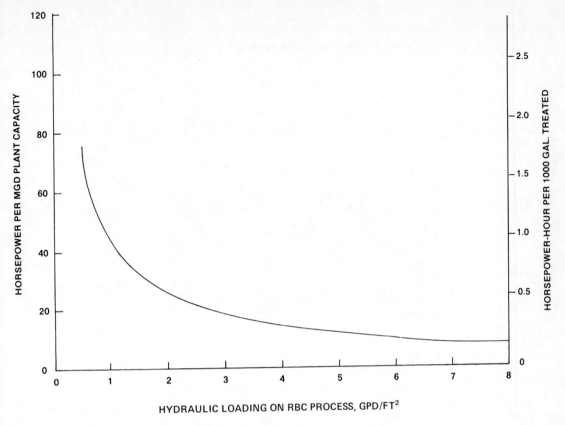

FIGURE 7. RBC process power consumption.

Figure 9-10 shows the present value of a single horsepower of energy saved as a function of power cost and interest rate for a 20-year period. For a power cost of 1.5¢/kWh and an interest rate of 6%, each horsepower saved in a treatment plant is worth $1,125. For even a medium-sized plant, about 3 mgd, the rotating contactor process can be expected to save 100 hp over activated sludge treatment. The present value analysis indicates that this is equivalent to a saving of $112,500 in initial capital cost.

Because municipalities are eligible for public funds for treatment plant construction, operating costs become even more significant in determining total treatment costs. If the medium-sized plant mentioned above is eligible for as much as 80% construction cost funding, the present value of the power cost savings becomes equivalent to 5 × $112,500 or $562,500 for the municipality.

Table 9-1 allows the determination of the present value of power and other operating cost savings for other time periods. To determine the

present value of an annual operating cost savings, simply multiply the annual savings by the factor listed in Table 9-1 for the appropriate period of time and prevailing rate of interest. For example, an annual operating cost saving of $100,000 accruing from savings in power consumption, operating personnel, chemicals, and other sources expected to continue for a period of 25 years at a prevailing rate of interest of 6% would be $100,000 × 12.783 = $1,278,300.

To adjust the present value of power savings determined from Figure 9-10 for time periods other than 20 years, simply multiply the indicated savings by the ratio of the factors indicated in Table 9-1. For example, from Figure 9-10 a savings of 1.0 hp at a power rate of 1.5¢/kWh and at an interest rate of 6% has a 20-year present value of $1,125. For a 30-year period under similar conditions, the present value would be $1,125 × $\frac{13.765}{11.470}$ = $1,350; for a 10-year period under similar conditions, the present value would be $1,125 × $\frac{7.360}{11.470}$ = $723.

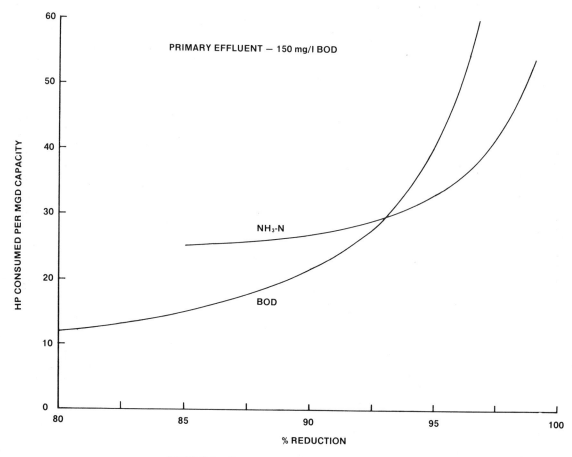

FIGURE 8.   Power consumption – primary effluent.

Total Cost Savings Analysis

Another technique for evaluating power savings is the annual cost saving. It uses the full value of a future flow of savings without discounting them back to the present. It is essentially a present value analysis at a 0% interest rate.

Total power savings are shown for several periods of time and as a function of power cost in Figure 9-11. At a power cost of 2¢/kWh, each horsepower saved is worth $4,670 in 30 years. For the example, of 100-hp savings mentioned above, this is worth $467,000.

Power costs certainly will rise significantly over the next 30 years. This will be especially true for large users who may have to face progressive rather than regressive rate schedules. If we conservatively assume the average cost of power at 3¢/kWh, the total cost savings is $700,000.

When the savings discussed above are compared to the RBC system costs in Figures 9-1, 2, and 3, it can be seen that for a 3-mgd plant, they represent

a significant portion of the total construction cost. This was also pointed out in a comparison of RBC and activated sludge designs by Winkler and Welch.[4]

COMPARISON WITH ACTIVATED SLUDGE

A comparison of total construction cost for a rotating biological contactor plant with an activated sludge plant will vary from application to application. For new plant construction, the costs shown in Figures 9-1, 2, and 3 will be approximately the same as for the activated sludge system, when it includes the aeration tank with aeration equipment and sludge recirculation and sludge wasting control systems. Because the activated sludge process produces a dilute waste sludge, a thickener and/or an enlarged digester are required for solids handling. This often results in a lower cost for the RBC process, especially for small- and

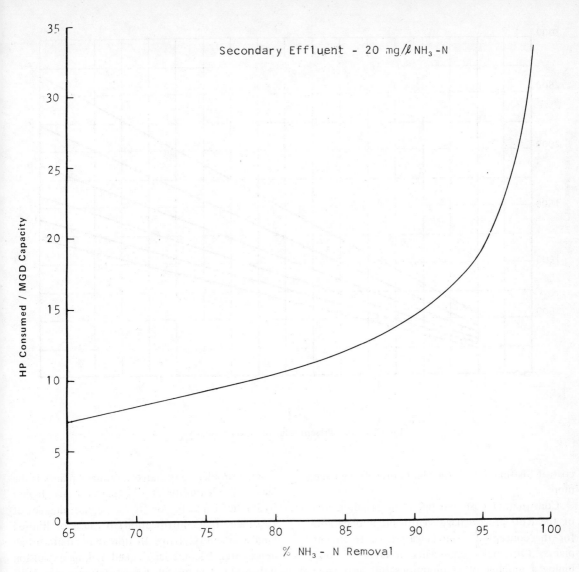

FIGURE 9. Power consumption — secondary effluent.

medium-size plants (1.0 to 10 mgd). Recent reports by the U.S. General Accounting Office[5,6] indicate that 40% savings in construction costs can be realized when substituting the RBC process for the activated sludge process in a 4.5-mgd treatment plant. While this amount of savings will not always be realized, it does point out that significant savings are possible and that the RBC process should be considered in all cost effectiveness studies. An even greater cost difference exists when a plant must be designed to achieve both BOD removal and nitrification where an activated sludge system must often be constructed in two stages with separate aeration settling and recirculation steps.

For very large plants, the activated sludge process is able to realize some economies of scale by using very large aeration tanks. The RBC process requires a large number of modular units so that economies of scale are not very significant for large plants. However, because the RBC portion of the plant represents about one third of the total plant cost, significant economies of scale are still available for the pretreatment, secondary clarification, and solids disposal portions of the plant costs. Under these conditions, the total construction cost of the two processes is about the same or slightly lower for activated sludge. Operating costs, however, for the RBC process are much lower, which usually result in a lower total

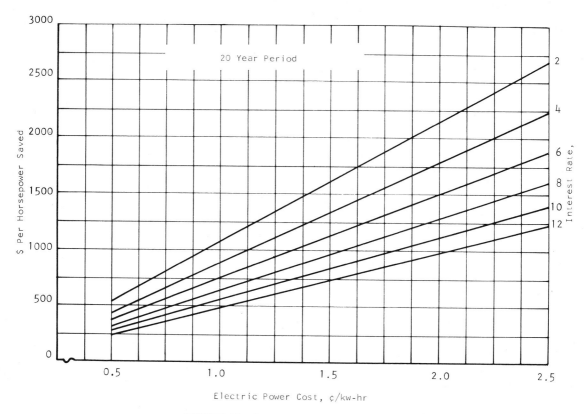

FIGURE 10. Present value of power savings.

cost of treatment even for plants of 100 mgd and more.

Construction cost and operating requirements of the RBC process have significant implications for the concept of regionalization of treatment plants. Two major arguments in favor of combining a number of communities into one large treatment plant are reduced construction cost from economies of scale that offset the high cost of additional sewer lines to connect them, and the improved plant operation by having one highly skilled central management and operations group rather than a number of smaller but necessarily equally skilled operations groups. The modular construction of the RBC system eliminates some of the economy of scale so the high cost of interconnecting sewer lines will often indicate that a number of smaller plants will be more economical. Because the RBC process requires relatively little low skilled attention, plant operation is no longer a major factor favoring regionalization of treatment plants. These factors must be kept in mind when making water resource management plans.

One problem with large treatment plants is the lack of flexibility for expansion to higher hydraulic capacity or to a higher degree of treatment (e.g., nitrification) because of limited land availability. Large treatment plants in urban areas often become landlocked making expansion unfeasible. The modular construction of the RBC process makes plant expansion quite simple for both small and large plants, whether the original plant was an RBC plant or used some other treatment process.

DIFFERENTIAL COST ANALYSIS

A comparison of costs between the RBC and activated sludge processes will only be meaningful when done for a specific application. Areas where differences in both capital and operating costs will occur for almost all applications will be pointed out here.

Pretreatment requirements for the two processes are similar but less critical for the RBC process. Pretreatment for the RBC requires removal of large and dense materials which may

TABLE 9-1

Present Value of One Dollar Received Annually At The End Of Each Year For N Years

Years (N)	1%	2%	4%	6%	8%	10%	12%	14%	15%	16%	18%	20%
1	0.990	0.980	0.962	0.943	0.926	0.909	0.893	0.877	0.870	0.862	0.847	0.833
2	1.970	1.942	1.886	1.833	1.783	1.736	1.690	1.647	1.626	1.605	1.566	1.528
3	2.941	2.884	2.775	2.673	2.577	2.487	2.402	2.322	2.283	2.246	2.174	2.106
4	3.902	3.808	3.630	3.465	3.312	3.170	3.037	2.914	2.855	2.798	2.690	2.589
5	4.853	4.713	4.452	4.212	3.993	3.791	3.605	3.433	3.352	3.274	3.127	2.991
6	5.795	5.601	5.242	4.917	4.623	4.355	4.111	3.889	3.784	3.685	3.498	3.326
7	6.728	6.472	6.002	5.582	5.206	4.868	4.564	4.288	4.160	4.039	3.812	3.605
8	7.652	7.325	6.733	6.210	5.747	5.335	4.968	4.639	4.487	4.344	4.078	3.837
9	8.566	8.162	7.435	6.802	6.247	5.759	5.328	4.946	4.772	4.607	4.303	4.031
10	9.471	8.983	8.111	7.360	6.710	6.145	5.650	5.216	5.019	4.833	4.494	4.192
11	10.368	9.787	8.760	7.887	7.139	6.495	5.988	5.453	5.234	5.029	4.656	4.327
12	11.255	10.575	9.385	8.384	7.536	6.814	6.194	5.660	5.421	5.197	4.793	4.439
13	12.134	11.343	9.986	8.853	7.904	7.103	6.424	5.842	5.583	5.342	4.910	4.533
14	13.004	12.106	10.563	9.295	8.244	7.367	6.628	6.002	5.724	5.468	5.008	4.611
15	13.865	12.849	11.118	9.712	8.559	7.606	6.811	6.142	5.847	5.575	5.092	4.675
20	18.046	16.351	13.590	11.470	9.818	8.514	7.469	6.623	6.259	5.929	5.353	4.870
25	22.023	19.523	15.622	12.783	10.675	9.077	7.843	6.873	6.464	6.097	5.467	4.948
30	25.808	22.396	17.292	13.765	11.258	9.427	8.055	7.003	6.566	6.177	5.517	4.979
40	32.835	27.355	19.793	15.046	11.925	9.779	8.244	7.105	6.642	6.234	5.548	4.997
50	39.196	31.424	21.482	15.762	12.234	9.915	8.304	7.133	6.661	6.246	5.554	4.999

From Anthony, R. N., *Management Accounting: Text and Cases,* Richard D. Irwin, Homewood, Illinois, 1960, 657. With permission.

otherwise settle in the rotating contactor tankage or plug the media. Any additional settleable material that may otherwise be removed in a conventional primary clarifier will pass through the system and settle in the final clarifier. In the activated sludge system, large and dense materials must also be removed so that they do not settle in the aeration tanks or become wrapped around aeration equipment. In addition, however, there is a desire to remove all relatively inert settleable matter. If not removed, these materials will pass through the aeration system and settle in the final clarifier just as in the RBC process. However, because solids are recycled in the activated sludge system, these inert materials will gradually accumulate. This will decrease the active fraction of the solids inventory in the process which will require higher MLSS in the aeration tank. This, in turn, will require more energy for mixing, result in a higher solids loading in the final clarifier, and require a higher sludge recycle ratio.

Tankage construction for the two processes is very different. The RBC process tankage is 5 to 6 ft deep and provides about 1 hr wastewater retention time for standard secondary treatment. Activated sludge tankage is usually 10 to 14 ft deep and provides 4 to 6 hr of wastewater retention time. This results in the RBC process occupying less land area. For some applications, land requirements may not be important. However, for plants in urban areas or for expansion of existing plants, especially if they are landlocked, this can be a critical consideration. When aeration tanks are made more than 20 ft deep, then the activated sludge process will occupy less land area. However, the energy required to mix and aerate the tank increases significantly so that this is usually not an attractive alternative. In either case, the cost of the tankage for the RBC system will be lower because of the smaller volume and shallow excavation. This cost difference becomes even greater when there is a high water table or when soil conditions require piling to support the weight of the tanks and wastewater. When tanks must be built above the ground, the reduced forming requirements and lower static water pressure signi-

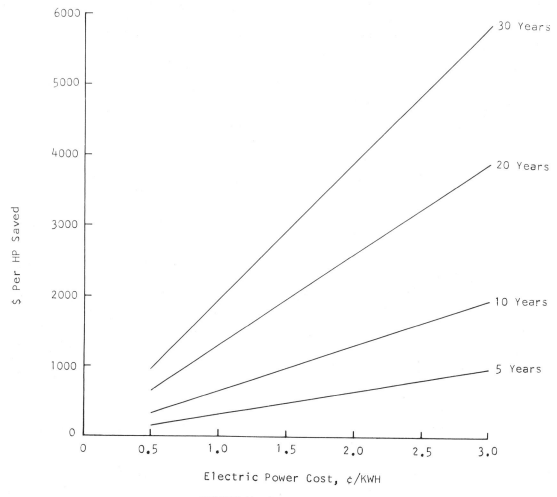

FIGURE 11.　Power cost savings.

ficantly reduce construction costs for the RBC process. Capital and operating costs for additional wastewater pumping are also reduced by the shallower tanks.

Aeration tanks are normally not covered. Because an enclosure is required for the RBC process, it is an additional cost. However, once covered, the RBC plant has a more aesthetic appearance, and provides better odor control and more operator comfort under extreme climatic conditions.

Another additional cost for the rotating contactor process on large plants is a large central installation of motor starters for each mechanical drive system. Use of the air drive system as described in Chapter 3 will eliminate most of this additional cost.

Secondary clarification is an area of significant cost savings potential for the RBC process.

Because surface overflow rate is the only important factor affecting solids separation efficiency, and because there is no sludge blanket in the bottom of the clarifier, the tankage need only be about 7 ft deep compared to 10 to 14 ft for the activated sludge process. This significantly reduces the cost of the basin and any piling needed to support it. The solids collection mechanism for the rotating contactor process should only be a simple chain and flight collector for rectangular tanks and rotating scraper blade collector for circular tanks rather than a suction or siphon-type needed for rapid sludge recirculation with the activated sludge process. Mechanical collectors are less expensive and simpler to operate.

Clarifier design overflow rates for the RBC process can be used for average daily flow conditions and not peak daily flow as is sometimes necessary for the activated sludge process. This is possible because of the low solids loading on the clarifier, which results in very little additional solids carry-over at peak flow. Although an over-flow rate of 800 gpd/ft^2 is recommended for most applications, there are many cases where much higher rates (1,200 to 2,000 gpd/ft^2) will give satisfactory effluent solids levels. This is true when treating dilute BOD concentrations where low solids concentrations are produced in the RBC effluent. In these cases, a relatively low solids separation efficiency will produce a satisfactory effluent. However, for the activated sludge process, a high solids separation efficiency is always required, because of the high solids loading on the clarifier.

Another significant area of cost savings for the RBC process is waste solids handling. A carefully operated clarifier following the rotating contactor process can produce an underflow solids concentration of 3%. This compares to 0.5 to 1.0% for the activated sludge process. When RBC solids are recycled back to a primary clarifier, a combined sludge of 4 to 5% is produced. While the mass of solids produced by the two processes will be approximately the same, the significantly lower volume produced by the RBC process results in lower solids handling costs. If thickening is to be used for solids dewatering, the rotating contactor process will significantly reduce or eliminate this cost. If sludge is to go directly to a digester, the RBC process will require a much smaller digester and will reduce the amount of energy needed to heat and mix the digester.

Operating cost reduction represents the greatest area of cost savings for the rotating biological contactor process. As discussed previously, energy conservation and energy costs play a significant role in determining total treatment costs. A direct comparison will be made with the activated sludge process to determine the extent of this savings potential.

Energy requirements for the activated sludge process are determined by the need for mixing and oxygen supply in the aeration tanks. For purposes of discussion this analysis will be done using diffused air. Smith[7] states that 1.0 SCF of air is required for 1 gal of wastewater to remove 117 mg/l of BOD from domestic waste. After allowing for a 5% transfer efficiency, Smith indicates an air requirement of 3,125 lb/hr or 1,120 ft^3/lb BOD at 68°F, and 8 psig discharge pressure. This provides about 0.75 lb oxygen/lb of BOD removed.

Other references indicate air rates as low as 600 to 700 ft^3/lb BOD for F/M ratios of 0.3 to 0.5,[8,9] and as high as 1,200 to 1,800 ft^3/lb BOD[9,10] for lower F/M ratios. The rate of 1,120 ft^3/lb BOD appears to be an appropriate value for this discussion.

Commercial blowers of the type generally used for wastewater treatment plants are 60 to 70% efficient. Listed in Table 9-2 are air rates/hp input for several centrifugal blower manufacturers. It is apparent that small blowers are less efficient than large blowers. This has a significant effect on power consumption of small vs. large activated sludge plants.

The air requirement of 1,120 CFM/lb BOD removed stated above was used for each influent BOD concentration in Figure 9-12. Effluent BOD values are based on the current definition of

TABLE 9-2

Centrifugal Blower Power Consumption
CFM/HP
At full capacity and 8.0 psig discharge pressure

Capacity, CFM	Hoffman[11]	Roots/Dresser[12]	Used for Analysis
100	9.1	17.5	13
500	13.2	–	15
1,000	16.7	–	18.5
2,000	18.2	–	19.5
5,000	20.4	24.3	22.5
8,000	21.0	24.2	23.0
10,000	20.8	24.3	23.0
12,000	–	25.0	23.0
15,000	–	24.6	23.0

secondary treatment, i.e., 30 mg/l effluent BOD or 85% reduction, whichever yields a lower effluent concentration. Power consumption for each case was determined using the figures in the right-hand column in Table 9-2. This yields a low value for power consumption, because it assumes that the blowers will be operated at full capacity, and normally they are somewhat oversized.

Figure 9-12 presents power consumption for an activated sludge process aeration system as a function of plant size and amount of BOD removed. Voegtle[13] states that actual power consumption can be 50% or more above the amount calculated, so the values shown in Figure 9-12 may be conservative.

For an RBC system of any size, the power consumption would be as shown in Table 9-3. Because the rotating contactor system is approximately first order with respect to BOD concentration, the power consumption per unit flow does not vary much as the BOD concentration varies. Comparing the above power consump-

tion for the RBC process with Figure 9-12 indicates significant power savings over the activated sludge process for all the treatment levels shown but especially for higher influent BOD values and smaller treatment plants. Power consumption for sludge recycle and air flotation thickening[7] was added to the activated sludge process requirements but not to the RBC process in Figure 9-13 for a typical influent BOD of 150 mg/l. This shows the differential power consumption for the two treatment systems. For a plant of 3 mgd capacity, about 100 hp of energy savings can be realized by using the RBC process rather than the activated sludge process. This analysis does not take into account any power savings realized from reduced secondary clarifier requirements or reduced solids handling requirements.

Another operating cost item often overlooked is the reduced manpower requirements for the RBC system. Because no process control functions are required of the operator, the level of skill necessary to operate an RBC plant is just that needed to provide adequate mechanical main-

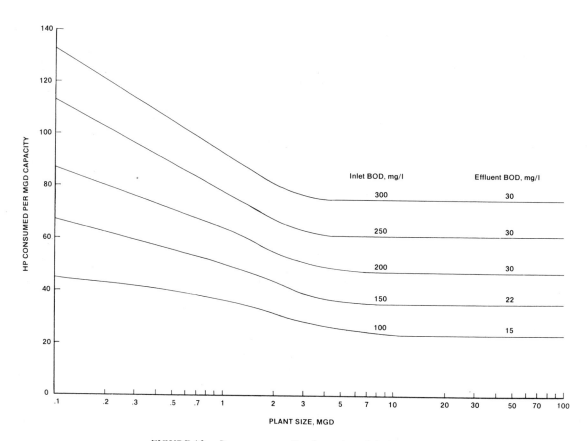

FIGURE 12.   Power consumption for activated sludge plants.

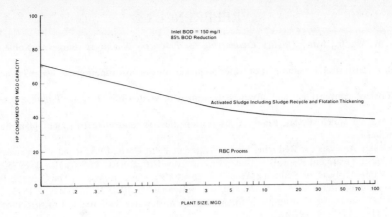

FIGURE 13.   Comparative power consumption.

TABLE 9-3

RBC Process Power Consumption

Inlet BOD mg/l	Effluent BOD mg/l	Hydraulic loading gpd/ft² (Figure 4-1)	Power consumed hp/mgd (Figure 9-7)
300	30	2.7	20
250	30	3.2	17
200	30	4.3	14
150	22	3.9	16
100	15	3.0	18

tenance to the drive components. Laboratory manpower is also reduced, because analytical requirements are limited to monitoring plant influent and effluent conditions. Because there is no need to control sludge age or determine the amount of sludge to waste or determine the sludge recycle rate, all the sampling and analytical work required for this function can be eliminated. With current labor costs of $25,000 to $50,000 per man year, this manpower reduction represents a significant savings in operating cost to a municipality or industry.

There has been considerable interest recently concerning the use of pure oxygen in conjunction with the activated sludge process. Many claims of capital and operating cost reductions over conventional activated sludge have been made. However, a recent comparison of pilot plant and full-scale experience using pure oxygen with conventional activated sludge by Kalinske[14] challenges most of the claims. The study concludes that there is no significant difference in capital or operating cost between air and pure oxygen activated sludge. Therefore, the cost comparisons and differential analyses presented here for RBC vs. activated sludge can also be considered valid for RBC vs. pure oxygen activated sludge.

The information presented in this chapter was not meant to imply that the RBC process will always be the lowest cost system for all applications. Specific applications will vary significantly so that generalizations regarding treatment process and system costs will not be valid. However, it is apparent that the RBC process offers significant cost savings and energy conservation potential and therefore should be considered in all cost effectiveness studies for treatment process selection.

REFERENCES

1. **Eckenfelder, W. W., Jr.,** *Water Quality Engineering for Practicing Engineers,* Barnes & Noble Books, New York, 1970, 292.
2. **Tihansky, D. P.,** Historical development of water pollution control cost functions, *J. Water Pollut. Control Fed.,* 46, 813, 1974.
3. **Smith, R.,** Cost of conventional and advanced treatment of wastewater, *J. Water Pollut. Control Fed.,* 40, 1546, 1968.
4. **Winkler, W. W. and Welch, F. M.,** Energy conservation dictates innovative treatment plant design, *Public Works Mag.,* March 1974.
5. Potential of Value Analysis for Reducing Waste Treatment Plant Costs, EPA, Results of Value Analysis Workshop Studies of A Waste Treatment Plant, U.S. General Accounting Office, RED-75-367, May 8, 1975.
6. Potential of Value Analysis for Reducing Waste Treatment Plant Costs, Report to The Congress by The Comptroller General of The United States, EPA, U.S. General Accounting Office, RED-75-367, May 8, 1975.
7. **Smith, R.,** Electrical Power Consumption for Municipal Wastewater Treatment, Environ. Protect. Technol. Ser., EPA, R2-73-281, p. 17 July, 1973.
8. Sewage treatment plant design, WPCF Manual of Practice No. 8, ASCE Manual of Engineering Practice, No. 36, p. 135, New York, N.Y., 1959.
9. **Fair, M., Geyer, J., and Okun, D.,** *Water and Wastewater Engineering,* Vol. 2, John Wiley & Sons, New York, 1968, 35.
10. Recommended Standards for Sewage Works (Ten States Standards), Great Lakes – Upper Mississippi River Board of State Sanitary Engineers, Health Education Service, Albany, N. Y., 1968, p. 80.
11. Quiet Power from Hoffman Centrifugal Blowers and Exhausters, Hoffman Air Systems Division of Clarkson Industries, Inc., ACB-898B/11M70, p. 10, 1968.
12. Roots Type MV Multistage Centrifugal Compressors, Industrial Products Division, Dresser Industries, Inc., Form No. B-5802.
13. **Voegtle, J. A.,** Be conservative about energy, Deeds and data, *Water Pollut. Control Fed.,* Feb. 3, 1975.
14. **Kalinske, A. A.,** Comparison of Air and Oxygen Activated Sludge Systems, paper presented at the 48th Annu. Conf. Water Pollut. Control Fed., Oct. 5-10, 1975, Miami Beach, Fla.

Chapter 10

CONCLUSION

The information that has been presented in this book represents the state of the art on the rotating biological contactor process in 1975. It is not intended to imply that all aspects of the process and its applications have been thoroughly investigated or optimized. Additional information from research and field experience continue to accumulate and will both enhance and modify the conclusions presented here.

Just a few of the many areas requiring additional work to fully optimize the process and make use of its favorable characteristics include optimization of rotational speed in successive stages of treatment, study of high rate solids separation and solids dewatering techniques in conjunction with the RBC process, response of the process to transient conditions particularly for nitrification applications, design considerations for scale-up from pilot plant data to full-scale commercial operation, additional testing on denitrification techniques with the RBC, and developing application data on a broader spectrum of industrial wastes particularly in the chemical processing industry. Some of these subject areas, particularly the response to transient conditions, can be studied efficiently through the development of mathematical models. To date, this has been done only on a limited basis and with very simple models. Work is currently being conducted in each of these areas and in the near future will provide a basis for revision and expansion of the state of the art as presented in this book.

APPENDIX

Additional References on the Rotating Biological Contactor Process Not Previously Cited

1. A Pilot Plant Investigation of Rotating Biological Surface Treatment of Pulp and Paper Wastes, National Council of the Paper Industry for Air and Stream Improvement, Stream Improvement Tech. Bull. 278, Nov. 1974.

2. T. W. Beak Consultants, Ltd., An Evaluation of European Experience with the Rotating Biological Contactor, Water Pollut. Control Directorate Environ. Protect. Services, Canada, Rep. No. EPS4-WP-73-4, Oct. 1973.

3. Antonie, R. L. and Van Aacken, K., Rotating discs fulfill dual wastewater role, *Water Wastes Eng.,* Jan. 1971.

4. Antonie, R. L., Rotating discs provide flexible treatment, *Ind. Water Eng.,* Aug./Sept. 1972.

5. Antonie, R. L., The Bio-disc process: New technology for the treatment of biodegradable industrial wastewater, *Chem. Eng. Symp. Ser., Water—1970,* 64(107), 585, 1970.

6. Borchardt, J. A., Observations on the Rotating Biological Discs, paper presented at the American Chemical Society Meeting, Chicago, Ill., 1970.

7. Bretscher, U., Phosphate elimination with RBC's, *GMF Das Gas-Und Wasserfach,* 110(20), 538, 1969.

8. Bringmann, G. and Kucha, R., Biological degradation of mineral oil products with the Bio-disc process, *Gesund. Singenienr,* 89(6), 179, 1968.

9. Burm, R., Cochrane, M., and Dostal, K., Cannery Wastewater Treatment with RBC and Extended Aeration Pilot Plants, Proc. 2nd Natl. Symp. Food Processing Wastes, March 23—26, 1971, Denver, Colo.

10. Chittenden, J. and Wells, W., Jr., Rotating biological contactors following anaerobic lagoons, *J. Water Pollut. Control Fed.,* 43, 746, 1971.

11. Clagget, F. G., Secondary treatment of salmon canning waste by rotating biological contactor (RBC), *Fish. Res. Board Can. Tech. Rep.,* Jan., 366, 1973.

12. Cochrane, M. and Dostal, K., RBC Treatment of Simulated Potato Processing Wastes, Proc. 3rd Natl. Symp. Food Processing Wastes, March 1972, New Orleans, La.

13. Combined Sewer Overflow Treatment by the Rotating Biological Contactor Process, EPA Rep., EPA-670/2-74-050 and Natl. Tech. Information Services, U.S. Dept. of Commerce, PB-231 892, June 1974.

14. Continued Study of Advanced Waste Treatment Systems for Combined Municipal and Pulp and Paper Wastes, The Institute of Paper Chemistry, Project 3029 — Rep. 5, July 10, 1974.

15. Corneille, R. W., Treatment of Apple Wastes Using Rotating Biological Contactors, paper presented at the 30th Purdue Ind. Waste Conf., May 6—8, 1975, W. Lafayette, Ind.

16. Duncan, P. and Walski, T., The Rotating Biological Contactor — Process Description, Mathematical Models, Design Applications, M.S. thesis, Vanderbilt University, Nashville, Tenn., 1973.

16a. Hao, O. and Hendricks, G., Rotating biological reactors remove nutrients, *Water Sewage Works,* Part 1, 122, 70, 1975; Part 2, 123, 48, 1975.

17. Hoehn, P. and Ray, A., Effects of thickness on bacterial film, *J. Water Pollut. Control Fed.,* 45(11), 2302, 1973.

18. Jewell, W., Davis, H., Johndrew, O., Jr., Loehr, R., Siderewicz, W., and Zall, R., Egg Breaking and Processing Waste Control and Treatment, Environ. Protect. Technol. Ser., EPA-660/2-75-019, June 1975; paper presented at the 30th Purdue Ind. Waste Conf., May 6—8, 1975, W. Lafayette, Ind.

19. Joost, R. H., Systemation in Using the Rotating Biological Surface (RBS) Waste Treatment Process, Proc. 24th Purdue Ind. Waste Conf., May 6—8, 1969, W. Lafayette, Ind., 365.

20. Kolbe, F. F., A promising new unit for sewage treatment, *Die Sivifle Ingenieur Sud Afr.,* Dec., 327, 1965.

21. Kornegay, B. and Andrews, J., Kinetics of fixed film biological reactors, Proc. 22nd Purdue Ind. Waste Conf., May 6—8, 1968, W. Lafayette, Ind., 680; *J. Water Pollut. Control Fed.,* 4(11), R460, 1968.

22. LaGregg, M., Klippel, R., and Nemerow, N., An Industrial Waste Case History: The Animal Glue Industry, paper presented at the 5th Mid-Atlantic Ind. Waste Conf., Drexel University, November 1971, Philadelphia, Pa.

23. Len-Hing, C., Obayashi, A., Zenz, P., Washington, B., and Sawyer, B., Biological Nitrification of a High Ammonia Content Sludge Supernatant Under Ambient Summer and Winter Conditions by Use of Rotating Discs, The Metropolitan Sanitary District of Greater Chicago, Dept. of Res. and Dev., Rep. No. 75-3, Feb. 1975.

24. McAliley, J. E., A pilot plant study of a rotating biological surface for secondary treatment of unbleached kraft mill waste, *Tappi,* 57(9), 106, 1974.

25. Moore, J., Hegg, R., and Larson, R., Treatment of Beef Waste by a Rotating Biological Contactor, paper presented at the 6th Natl. Agric. Waste Manage. Conf., Cornell University, March 1974, Ithaca, N.Y.

26. Municipal Sewage Treatment with a Rotating Biological Contactor, Allis-Chalmers Research Center for FWQA, Contract 14-12-24, Mod. No. 2, May 1969, 17050 DAM 05/69, Natl. Tech. Information Service, W71-1655, PB 201 701.

27. Person, H. and Miner, J., An evaluation of three hydraulic manure transport treatment systems, including a rotating biological contactor, lagoons, and surface aerators, Journal paper no. 5-7152 of the Iowa Agric. and Home Economics Equipment Stations, Ames, Ia., Project No. 1730; paper presented at the 1972 Cornell Agric. Waste Manage. Conf., Syracuse, N.Y., Jan. 31, 1972.

28. Pretorius, W. A., The Complete Treatment of Raw Sewage with Special Emphasis on Nitrogen Removal, paper presented at the 6th Int. Conf. Adv. Water Pollut. Res., June 8—23, 1972, Jerusalem.

29. **Roskopf, P., Osborn, F., Watson, D., and Flann, G.,** Rotating Biological Surface Treatment of Vegetable Canning Process Wastewater, paper presented at the 6th Natl. Symp. on Food Processing Wastes, April 11, 1975, Madison, Wis.

30. Rotating Biological Contactor, Allis-Chalmers Research Center for FWQA, Contract 14-12-24, Phase 1 Rep., 1969.

31. **Storer, E. and Kincannon, D.,** One-step nitrification and carbon removal, *Water Sewage Works,* 122, 66, 1975.

32. **Summer, R. and Bennett, D.,** Effluent treatment by rotating biological surface, *Pap. Trade J.,* 157, 60, 1973.

33. **Tchobanoglous, G.,** Wastewater treatment for small communities, *Public Works,* Part 1, 105, 61, 1974; Part 2, 106, 58, 1974.

34. **Thomas, J. and Koehrsen, L.,** Activated Sludge – Bio Disc Treatment of Distillery Wastewater, Environ. Protect. Technol. Ser., EPA-660/2-74-014, April 1974.

35. **Torpey, W., Heukelekian, H., Kaplovsky, J., and Epstein, R.,** Effects of Exposing Slimes on Rotating Discs to Atmosphere Enriched with Oxygen, paper presented at the 6th Int. Conf. on Water Pollut. Res., June 18–23, 1972, Jerusalem.

36. **Willard, H., Eckerle, W., and Scott, R.,** Feasibility of Rotating Disc Treatment Process for Hardboard and Insulation Board Wastewater, paper presented at the Forest Products Res. Assoc. Meeting, June 21, 1972, Dallas, Tex.

37. **Wilson, R., Murphy, K., Sutton, P., and Jank, B.,** Nitrogen Control: Design Considerations for Supported Growth Systems, paper presented at the 48th Annu. Water Pollut. Control Fed. Conf., Oct. 5–10, 1975, Miami Beach, Fla.

INDEX

A

Actifil
  as trickling filter media, 3
Activated sludge process
  capital cost, 180–182
    vs. RBC, 182–185
  development, 1
  flexibility, 12
  maintenance, 12
  nitrification, 12
  operating cost, 181–182
    vs. RBC, 185–187
  power consumption, 12
  retention time, 12
  stability, 12
  upgrading, 81–91
    denitrification, 91
    design procedure, 86–90
    gravity filtration of effluent, 85
    pilot plant tests, 84–86
  with trickling filter, 3
Aeration
  BOD removal, 130
  RBC vs. trickling filter, 11
Aerobic digestion
  of secondary sludge, 65
Aerobic pretreatment
  domestic wastewater, 68
Air drive system
  advantages of, 44
Allis-Chalmers, 11
Allythiourea
  as additive to dilution water, 48
  in BOD removal, 21
Alum
  effect on efficiency, 155
  in phosphate removal, 65–66
  in phosphorus removal, 153
Aluminum
  disc, 37
American Distilling Company, 169–170
Ammonia
  oxidation, 35
Ammonia nitrogen
  as primary effluent, 16
  concentration
    effect on nitrification, 57
  denitrification, 64
  effluent concentration, 27–30, 34, 37
    relation to Kjeldahl nitrogen, 144
  in refinery waste, 114
  oxidation, 28
  removal, 28, 34, 36, 37, 144
    as function of retention time, 23
    domestic wastewater, 53
    effect of staging, 17, 19
    in two-stage pilot plant, 17

B

Ammonium hydroxide
  in hydraulic surge, 97
Anaerobiasis
  effect on efficiency, 134
Anaerobic digester
  supernatant, 67
Analysis
  of wastewater, 135
Animal glue waste
  design criteria, 119
Asbestos
  enclosure, 128
Automatic sampling, 135
Autotral Corporation
  polystyrene disc, 11

Bacteria
  in media, 9
  in multistage test unit, 24
  white biomass, 132
Baffle
  construction, 124
  placement, 125
Bakery waste, 167–168
Beggiatoa, 132
Beverage waste
  design criteria, 106
BIO-DISC, 11
Biological wheel, 10
Biomass
  nature of, 7–9
BIO-SURF, 11
BOD
  as primary effluent, 16
  concentration
    effect of disc spacing, 28
    effect on nitrification, 38, 57
    in design calculation, 51
  effluent characteristics, 36
    as design criterion, 48
  removal
    as function of hydraulic loading, 34, 35
    as function of retention time, 22
    effect of disc velocity, 19
    effect of enriched oxygen atmosphere, 40
    effect of staging, 17
    effect of wastewater temperature, 32
    four- vs. six-stage operation, 31
    hydraulic loading, 20
    in industrial wastewater, 99–102
    six-stage operation, 30
Boise Cascade Company, 171
Brewster treatment plant, 158–160

193

of industrial wastewater, 96–98
test system, 40
Hydrogen sulfide
effect on efficiency, 134
Hydrolysis
in nitrogen removal, 36

I

Imhoff tank
as primary clarifier, 65
domestic wastewater treatment, 68
upgrading, 77
Individual home treatment plants, 74
animal glue waste, 119
Industrial wastewater treatment
bakery waste, 167–168
beverage waste, 106
coal mine waste, 119
combined with domestic, 118–119
dairy waste, 106
design criteria
BOD removal kinetics, 99–102
enlarged first stage, 104–105
intermediate clarification, 102–104
pilot plant tests, 105–106
wastewater temperature, 105
distillery waste, 169–170
electronics manufacturing waste, 120
food-processing waste, 106, 163–167
latex polymer waste, 120
meat waste, 106
nylon manufacturing waste, 170
paper waste, 106–113, 170–172
poultry processing waste, 170
pretreatment, 93–94
pulp waste, 106–113, 170–172
refinery waste, 113–118
textile waste, 119
variable wastewater flow
hydraulic surge, 96–98
intermittent, 94–96
organic surge, 96–98
winery waste, 168–169
Inorganic nitrogen
in industrial wastewater, 94
Intermediate clarification
equipment configuration, 128–129
of industrial wastewater, 102–104

K

Kinetic evaluation
of activated sludge process, 47
Kjeldahl nitrogen
as primary effluent, 16
denitrification, 64
effluent

relation to ammonia nitrogen, 144
removal, 35–36
Koroseal
as trickling filter media, 3
Kralis Poultry Company, 170

L

Latex polymer waste
design criteria, 120
Lagoon
domestic wastewater, 66
supernatant, 67
upgrading, 91–92
Loading
hydraulic, see Hydraulic loading
organic, see Organic loading

M

Maintenance
municipal treatment plant, 150
of equipment, 135–136
preventive, 136
RBC vs. activated sludge process, 12
Manpower
cost, 186–187
Manual sampling, 135
Meat waste
design criteria, 106
Media
construction, 32–33
corrugated
schematic drawing, 33
corrugated polyethylene, 41
high density
cost, 175–176
in concrete tankage, 72
rotation
as design criterion, 48
Methanol
as carbon source, 53
denitrification with, 91
Microscreening
of domestic wastewater, 66
Mixed liquor
characteristics, 22–23, 26
effect of sludge recycle, 23
in municipal wastewater, 146–150
Municipal wastewater treatment, see also Pilot plant
demonstration plant, 137
0.50-mgd plant
effluent characteristics, 143–146
maintenance, 150
mixed liquor characteristics, 146–150
operation, 150
sludge characteristics, 146–150
tertiary filtration, 150

S

Sampling
 of wastewater, 135
 stage analysis, 155–157
Secondary clarification
 cost, 184
 domestic wastewater, 64
 separation of solids, 9
 sludge characteristics, 12
Secondary treatment
 municipal wastewater, 151–154
Septic tank
 in sludge disposal, 72
 over-and-under configuration, 72
 pretreatment, 68–70
 upgrading, 77
Settling
 in nitrogen removal, 36
Shaft assembly
 alternative drive system, 43–44
 large plant, 124–127
 small plant, 123–124
Shearing force
 control of biomass, 34
 removal of biomass, 7
Side-by-side configuration, 70–71
 Brewster treatment plant, 158
 key features, 72
 with media assembly in concrete tankage, 73
Six-stage operation, 30
Slime
 in trickling filter, 7
Sloughing, 131
 controlled
  RBC vs. trickling filter, 11
 ineffective
  RBC vs. trickling filter, 11
Sludge
 anaerobic digester, 65
 disposal, 64, 72
 handling, 64
 in municipal wastewater, 146–150
 production
  as function of degree of treatment, 25
  calculation, 22
  in municipal wastewater, 154
 recycle, 23–24, 134–135
  effect on treatment efficiency, 27
 treatment supernatant, 66
Sludge age
 effect of temperature, 58
Sludge scoop
 in clarification, 72
Sphaerotilus
 in multistage test unit, 24
Stability
 RBC vs. activated sludge process, 12
Staging, 30
 as design criterion, 47–48
 in upgrading activated sludge system, 90

large plant, 125–126
 two vs. four, 17–19
Subadjacent clarification, 77–78
 equipment configuration, 128
Sulfide ion, 121
Surfpac
 as trickling filter media, 3
Suspended solid
 as primary effluent, 16
 effluent, 36
  effect of disc spacing, 29
 removal
  as function of hydraulic loading, 34, 35
  effect of disc spacing, 19, 29
  in design calculation, 53

T

Tank volume
 as design criterion, 48
Teen Challenge Training Center, 160
Telescoping valve
 efficiency, 134
Temperature, see also Wastewater temperature
 correction
  in design calculation, 51
  in nitrogen control, 57–60
 effect on efficiency, 132, 139, 143, 146–149
 effect on industrial wastewater treatment, 105
 effect on nitrification, 90
 effect on sludge production, 154
 effect on upgrading lagoon, 92
 of wastewater
  as design criterion, 49
Tertiary filtration
 municipal wastewater, 150
Tertiary treatment
 domestic wastewater, 66
Textile waste
 design criteria, 119
Thermal conditioning, 66
Thermal insulation
 enclosure, 128
Thio salts, 121
Thiotrix, 132
Total cost savings analysis
 power consumption, 180
Toxicity reduction
 in paper wastewater, 112
 in pulp wastewater, 112
 of refinery waste, 116
Transportable plants, 74
Trickling filter, 1–5
 cost, 4–5
 decline of use, 3
 design criteria
  hydraulic loading, 4
  organic loading, 4
 economic problems, 2
 fundamentals, 1